Yearbook of the Artificial

Yearbook of the Artificial | Volume 3

Edited by
Massimo Negrotti

In the different areas of the artificial – defined as the attempt to reproduce natural objects or processes by means of the available technology – researchers and designers often work within their disciplines with little or no information about research being carried out in other areas. Bioengineers, for instance, know little about the problems of Artificial Intelligence researchers, while roboticians often neglect the efforts of medical engineering, etc. The Yearbook of the Artificial discusses current theories, projects and models from a wide range of areas without going into too many technical details. The main assumption is that there are common logical and methodological problems for all disciplines that are concerned with the reproduction of natural objects and processes. The Yearbook provides a forum for all of them, and covers areas such as bioengineering, robotics, A.I., artificial neural networks, artificial life, multimedia, the history of technology, communications, art and music.

In submitting a paper to the *Yearbook of the Artificial*, you grant us permission to publish that paper after a peer review. We ask that your paper be between 20000 and 24000 characters, and pertinent to the topics mentioned above. We also require an abstract of no more than 150 words, including a list of keywords. Papers should be submitted in triplicate, typewritten, double-spaced, on one side of the paper. Please include a copy on a floppy disk in a commonly used word processor format.

Only original papers in English should be sent to:

Dr. Domenico Donnanno
dodo@soc.uniurb.it
IMES-LCA
Via Saffi, 15
University of Urbino
I-61029 Urbino (Pu) Italy

Yearbook of the Artificial

Nature, Culture & Technology

Vol. 3

Cultural Dimensions of the User

Edited by Massimo Negrotti

PETER LANG
Bern · Berlin · Bruxelles · Frankfurt a.M. · New York · Oxford · Wien

Bibliographical information published by Die Deutsche Bibliothek
Die Deutsche Bibliothek lists this publication in the Deutsche Nationalbibliografie; detailed bibliographical information is available on the Internet at ‹http://dnb.ddb.de›.

British Library and Library of Congress Cataloguing-in-Publication Data:
A catalogue record for this book is available from *The British Library,* Great Britain, and from *The Library of Congress,* USA.

Il presente volume è pubblicato nell'ambito della Ricerca 'Modelli dell'utente nelle società avanzate', finanziata dal M.I.U.R. e dall'Università degli Studi di Urbino 'Carlo Bo', COFIN 2002, Prot. n. 2002143319.

This volume is published as part of the research project 'Models of the User in Advanced Societies', co-financed by the Italian Ministry of University Teaching and Scientific Research (M.I.U.R.) and by the University of Urbino 'Carlo Bo', COFIN 2002, Prot. n. 2002143319.

Cover image: Massimo Negrotti
Cover design: Thomas Jaberg, Peter Lang AG

ISSN 1660-1084
ISBN 3-03910-690-2
US-ISBN 0-8204-7556-4

Hochfeldstrasse 32, Postfach 746, CH-3000 Bern 9, Switzerland
info@peterlang.com, www.peterlang.com, www.peterlang.net

Scientific Committee

Ashok, Jain
Professor of Computer Science
Institute of Informatics and Communication
Delhi University South Campus, New Delhi, India

Baggi, Denis L.
Head, Dept. of Communication and Multimedia
Institute CIMSI of the Professional University of Southern Switzerland
Manno, Switzerland

Bertasio, Danila
Professor of Sociology of Art, IMES-LCA
University of Urbino, Italy

Borillo, Mario
Research Director at the CNRS, Emeritus
Institut de Recherche en Informatique de Toulouse
Université Paul-Sabatier
Toulouse, France

Chang, Thomas
Director, Artificial Cells & Organs Research Centre
Professor of Physiology, Medicine & Biomedical Engineering
Faculty of Medicine, McGill University,
Montreal, Canada

Cordeschi, Roberto
Professor of Philosophy of Science
Dept. of Communication Sciences
University of Salerno, Italy

Fantini, Bernardino
Professor of History of Medicine
Institute of History of Medecine
Faculty of Medicine
University of Geneva, Switzerland

Gill, Karamjit S.
Professor of Human Centred Systems
School of Information Management
University of Brighton, UK

Honavar, Vasant
Professor of Computer Science
Department of Computer Science
Iowa State University, Ames, USA

Lanzavecchia, Giuseppe
Visiting Professor of Sociology of Science, IMES-LCA
University of Urbino, Italy

Longo, Giuseppe O.
Professor of Information Theory
Department of Electrical and Electronics Engineering
University of Trieste, Italy

Marchis, Vittorio
Professor of Applied Mechanics and History of Technology
Department of Mechanics
Polytechnic of Turin, Italy

Margolin, Victor
Professor of Art and Design History
Dept. of Art History
University of Illinois, Chicago, USA

Negrotti, Massimo
Professor of Methodology of Human Sciences, IMES-LCA
University of Urbino, Italy

Nuki, Takao
Professor of Business Administration
Dept. of Management
Musashi University, Tokyo, Japan

Nissan, Ephraim
Research Fellow in Artificial Intelligence
School of Computing and Mathematical Sciences
University of Greenwich, Old Royal Naval College
Greenwich, London, England

Oliverio, Alberto
Professor of Psychobiology
Dept. of Genetics and Molecular Biology
University of Rome "La Sapienza"
Rome, Italy

Qvortrup, Lars
Professor of Interactive Media
Department of Interactive Media
University of Southern Denmark, Odense, Denmark

Rizzotti, Martino †
Professor of Evolutionary Biology
Department of Biology, University of Padova, Italy

Sato, Yoshihiro
President, Ph.D.
Institute for Culture and Information Technology, Inc., Tokyo, Japan

Satofuka, Fumihiko
Professor of History of Science
Department of Eco-Design
Musashino Women's University, Tokyo, Japan

Trautteur, Giuseppe
Professor of Cybernetics, Department of Physics
University of Naples, Italy

Technical secretaries:

Dr. Domenico Donnanno
Dr. Julian R. Locke

Table of Contents

MASSIMO NEGROTTI
Humans and naturoids: from use to partnerships 9

DANILA BERTASIO
Between seductions and banalities:
the case of art on the web 17

ELENA BETTINELLI
Religions on the Internet: a comparison
between different kinds of websites and users 33

MONICA BORDEGONI, UMBERTO CUGINI
AND ROSSELLA MANA
Novel interaction methods exploit designers' skills 47

LUCIANO CELI
'It's like if': the power of exhibits in Science Centres
and Museums between immersion and illusion 65

GIANUGO M. COSSI
The user of contemporary art 83

FRANCO FILENI
Licence of use: the new owners of knowledge 101

GIUSEPPE LANZAVECCHIA
A conceivable model for user: fields and attractors 113

JULIAN LOCKE
On-line machine-based natural-language translation:
a new approach 123

RICCARDO MANZOTTI AND VINCENZO TAGLIASCO
Towards an artificial user: the 'what' problem
for an architecture capable of developing new goals 133

LELLA MAZZOLI, ROBERTA BARTOLETTI AND GIOVANNI BOCCIA ARTIERI
The artificial in everyday life:
social uses of communication technologies 147

SABRINA MORETTI
Citizens, users, society: a sociological survey 157

TERESA NUMERICO AND ROBERTO CORDESCHI
Meno's dilemma and web mining: the influence
of search engines on the discovery process 175

GIACOMO ROMANO
The problem of the individuation of artifacts 189

DAVID SMITH
Fly's Eye View: an investigation of the impact of the
experience of using a VR system on young children's
understanding of representation and reality 203

MARISELDA TESSAROLO
Transition of the relation between artist and user
to that between culture supply and demand 215

LAURA VERDI
The places of the users 225

MARIA ROMANA ZORINO
The client-assigner of cultural artifacts 239

Massimo Negrotti

Humans and naturoids: from use to partnerships

Abstract:

We all know that something can be used if it cannot do anything by itself, but stays in some standby state or, at maximum, self-regulates some of its states according to some established rule.

This means that tools and machines can be said to be used if and only if they cannot take any initiative as autonomous entities. By contrast, a true relationship is always to be conceived as a peer-based interaction, wherein each partner can start some action unpredicted by the other. All this introduces peculiar problems when having to deal with devices, such as advanced robots conceived as high-level naturoids, whose aim is just that of reproducing the modes natural systems follow when they take some unpredictable but meaningful initiative.

Keywords:
naturoids, autonomous machines, users, partnerships

1 Humans as users and enjoyers

The relationships between man and machine are at the core of contemporary research. Nevertheless, if we examine this issue by replacing the term 'man' with that of 'user' and the term 'machine' with that of 'naturoid' (or, at least, 'non-trivial machine', to adopt von Foerster's expression), then we will discover a lot of unexpected problems. In fact, if we adopt the term naturoid for referring to the whole set of the artefacts designed for reproducing natural exemplars [Negrotti, 2002, 2004], then its connection with the figure of the 'user', as someone who uses something, assumes a rather paradoxical character.

All this is clear even regarding automatisms, which are often nothing but a sort of *hidden* naturoids (self-regulating systems replacing human control). Consider the expression 'I use the automatic door'. Though it is very clear, it is easy to sense something contradictory in such a statement, because, if an artefact is automatic, then by definition it does not need any operation by the user; that is to say, he does not make any evaluations, regulations, etc. If, for example, when leaving a department store, we decide to avoid the traditional door, which truly implies the use of some artefact, such as a handle or a button, and choose instead the automatic door, we are no longer *users*, and we become, so to speak, *enjoyers*. In such circumstances, we should simply say that we are 'passing through the door' and not, strictly speaking, that we are 'using' it. To sum up, on the one hand we *use* tools, and on the other we *enjoy* automatisms. In the field of natural objects or systems this point is even clearer. Who would claim to be user of his own heart, or of a storm or a feeling or a concert?

In the field of naturoids – particularly the ones which, in contrast to automatisms, are machines we can readily locate and see in action as something intelligent – the incongruity we are speaking about emerges in a different way. However, if we examine it carefully, we discover that the situation is quite the same. Indeed, since a naturoid includes, generically, partly something natural and partly something technological, the adoption of the term *user* rather than the adoption of the term *enjoyer* (or something like that), indicates our propensity to define a given naturoid as something closer to nature than to technology, or vice versa.

This discussion does not involve a mere linguistic question. Rather, we have to understand how, thanks to the above-mentioned incongruities, one can discover more effectively the whole series of novelties that the advanced technologies, and those of naturoids in particular, are introducing in our relationships with the technological environment.

2 Can we 'use' a robot?

The key point resides in the degree of autonomy we are inclined to attribute, conceptually or *de facto*, to a naturoid. Therefore, the adoption of the verb 'to use' regarding a drill, which is not a naturoid, appears to be quite correct, while it is easy to perceive a certain inadequacy when adopting the same verb with regard to a robot. The inadequacy would be even more apparent if we had to deal with a highly flexible robot, able to regulate its own behaviour, and maybe designed to appear anthropomorphic. In other words, with such a robot – but we could extend this argument to a large part of naturoids – we really set up a relationship or, better to say, a partnership.

Actually, such a reality is already present in many human activities – even the more ancient and simple ones – provided they have at their centre objects, artefacts or events inspired by nature. No little girl, for instance, would say that she is *using* her doll: rather, she would say that she is playing with it. In the same way, nobody would define himself as a user of a painting that reproduces a landscape, nor of the landscape itself.

More generally, we can maintain that one will sense the incongruity of the expression 'to use' when one thinks of the artefact one has to deal with as something lacking any autonomy (and even less autopoiesis). On the contrary, when some form of autonomy seems to arise in the behaviour of the artefact, or when, for some reason, one is inclined to attribute some autonomy to it (as in the case of the doll, but also in many forms of totemism, animism, magic and so forth), then one perceives the emergence of a true relationship.

This means that a relationship implies the setting up of a system whose evolution, *ceteris paribus*, is not a deterministic one, in the sense that man is not the only player, and the object also has a more or less relevant ability to take the initiative. In fact, while a traditional man-machine system may easily be defined as a univocal *master-slave system*, when such a system includes a sufficiently developed naturoid or a truly non-trivial machine, it an open and uncertain relationship emerges, in which the roles of master and slave can swap very frequently.

3 The autonomy of naturoids and its acknowledgement

Beyond the purely linguistic question, therefore, the anthropological profile becomes important, because we lack any objective measure of autonomy (which should not be confused with so-called 'degrees of freedom') of a naturoid or machine. As a consequence, the perception of the relationship that is developing between a man and a naturoid will strongly depend on the way our culture develops in establishing some quality threshold beyond which we can accept that we are dealing with an autonomous entity.

For instance, although computers have dramatically improved their flexibility in their interactions with us, we find it hard to abandon the idea that, in the end, they are mere machines to be 'used'. On the other hand, several machines governed by one or more computers, possibly hidden, set up relationship-like situations that are in fact perceived as such, and not as simple use-like circumstances. We need only think of some kinds of particularly refined and powerful computer systems, of which Big Blue was one of the most remarkable recent instances. Even more impressive, however, are some advanced medical analysis devices from which we can expect diagnostic results that could not otherwise be obtained.

Nevertheless, apart from the world of computer-based naturoids, our disposition to perceive the relationship as one of a simple use greatly increases if we take into consideration other advanced fields, such as bioengineering.

All phenomenology of so-called artificial organs probably generates, in the human beings who adopt them, a very different set of effects when compared to the ones we may observe in whoever uses traditional tools or simple machines. In the case of artificial limbs, for example, the relationship is not only truly perceived as such, but can produce strong assimilation and appropriation effects: the organ, to sum up, becomes a part of us, and we rely on its autonomous capacity to support our overall survival. In such circumstances, humans reach the maximum of their disposition to confer dignity upon these objects, because of their autonomous ability to control what is going on in our body.

A further interesting case, then, is that of a special kind of *human-in-the-loop* system within which we try to attribute to the naturoid the ability to adapt itself to the behaviour of the human being with whom it is working. Here, a naturoid's autonomy (in evaluating and regulating its own behaviour) is wholly orientated towards the permanent attempt to comply with a human being's actions. Nevertheless, as has sometimes been observed, man, too, tries to adapt his own behaviour to the actions of the naturoid. In other words, both man and machine try to anticipate the evaluations and the actions of their partner. In so doing, they base their predictions on previous experience and on some model, built into their mind (or software), of the habits and expected behaviour of the partner.

All this triggers a potentially reflexive situation whose outcome can appear to be quite unforeseeable. The adaptation process we are speaking about, indeed, develops through more or less refined prediction techniques, thanks to which both man and machine predict the action of the partner. Moreover, both the partners may also be able to take into consideration the habits of the other in anticipating their own actions. Thus, a sort of *petitio principii* might begin, which, incidentally, reminds us of the logic of paradoxes.

Such processes, in all kind of relationships, generally find their solution through moments of intentional or unintentional passivity on the part of one of the partners, and this allows the system to converge to a steady state. If such a type of passivity should not emerge, then the system will tend towards a meaningless and an infinite oscillation, which, to date, is a very rare case in human systems, but is not so infrequent in man-machine systems.

Usually a situation of the latter kind generates frustration or even panic in a human being, while the naturoid or machine can continue to work indefinitely. This capacity to tolerate whatever kind of situation without human-like pains is, in the end, a major feature of current naturoids, artificial devices or non-trivial machines. This could imply that a man, as the physically weaker pole of the system, is potentially destined to succumb to the winner; that is to say, the naturoid or the machine has no totally autonomous target to reach, nor mysterious strategy to follow.

Stating it differently, it is quite possible that humans will gradually lower their will to control naturoids or machines, because of their different 'style' of living relationships, and because of their intrinsic complexity.

Thus, instead of trying always to understand what is going on in the relationships, humans would prefer to follow alternative strategies and relationships. This could be dangerous in some particular circumstances, of course. However, this in itself does not imply that the machines will take advantage of this fact to gain some kind of overall supremacy, as authors such as H. Moravec, B. Joy e R. Kurzweil are inclined to think [Moravec, 1989, 1998, Joy, 2000, Kurzweil, 2000]. Seemingly, there are no new masses of machines which push on our door and 'challenge our morality', as imagined by Latour. Rather, the real problem is that of finding general models for establishing an efficient and effective *symbiosis* between man and machines (culture and technology), a theme that has already gained some popularity in the Japanese debate on the *kyousei*. The concept of partnerships could be very useful in this direction, for it clearly indicates that advanced technology is not something we can *use* according to our old cultural models, in which master and slaves were openly recognizable.

Bibliography

K. S. Gill (ed.), *Human Machine Symbiosis: The Foundations of Human-Centred Systems Design*, Springer, London 1996

B. Joy, 'Why the future doesn't need us', in *WIRED*, 8, April, 2000.

R. Kurzweil, *The Age of Spiritual Machines: When Computers Exceed Human Intelligence*, Penguin Books, New York 2000

B. Latour, 'Where Are the Missing Masses: The Sociology of a Few Mundane Artefacts', in W. Bijker and J. Law (eds.), *Shaping Technology/Building Society: Studies in Sociotechnical Change*, MIT Press, Cambridge, Mass. and London 1992

V. Margolin, 'Human and Machine: Issues of Collaborative Action', in M. Negrotti (ed.), *Yearbook of the Artificial*, Vol. 1, Peter Lang, Bern 2002, pp. 107-118.

H. Moravec, *Robot: Mere Machine to Transcendent Mind*, Oxford University Press, Oxford 1998.

M. Negrotti, *Naturoids*: *On The Nature of the Artificial*, World Scientific Publishing Company, New Jersey 2002.

M. Negrotti, 'Naturoids: From Representations to Concrete Realizations', in *Pragmatics & Cognition*, John Benjamin, Amsterdam 2004, 12, 1, pp. 37-56.

Massimo Negrotti, LCA – Laboratory for the Culture of the Artificial, University of Urbino

Danila Bertasio

Between seductions and banalities: the case of art on the web

Abstract:

The aim of this work consists in trying to understand what effects the use of the Net can produce in the already fragile relationship between art and public. Will art on the Internet increase or decrease the demand, on the part of the public, to undertake a process leading to a mature aesthetic evaluation of the world and of the works of mankind? In spite of new technologies, which allow information to circulate much more easily than in the past, can we affirm that this has determined a consequent rise in critical abilities?

Focusing on the case of Italy in particular, this paper also sets out to examine the effects of recent political decisions regarding investment in, and promotion of, the arts. Have policies of substituting the public cultural budget with private money led to a restriction of the flow of resources to only the most proven values? Are cultural considerations being sidelined in favour of vested interests?

Finally, we ask: what future can we expect for art and for the art-going public?

Keywords:
artifacts, artifact individuation, artifact identification, artifact conceptualization, design stance, means-end reasoning, teleo-instrumental attitude

1 Premise

It is undeniable that the success of advanced technologies is, by and large, applicable to the fact that software and hardware, in the widest sense of the words, are always more accurately designed for a conveniently generic and indefinite public, which is perceived as being in need of the machine's services, independently from its degree of interest in technological issues.

The internal language of the machine, which allows us to establish an interactive relationship, becomes inaccessible; the architecture of software does not require or allow either any form of communication or any relationship between the person who installs a program and the one who uses it[1]. It is just not the case that nowadays, despite some newly born attempts, the group of prospects introduced in the 80s and 90s by the research tradition under the name of *Human Centred System*, within which was sustained the usefulness of involving users in the actual design of the software, is considered to be absolutely utopian [Gill, 1996]. The last twenty years have proved, in fact, the impracticability of such presuppositions, but not their theoretical fallacy. If, on the one hand, the inaccessibility of the systems has rendered the man-machine interaction more feasible, on the other hand, it has given rise to serious questions regarding the reliability and management of information.

Nowadays, History and Chronicle seem to converge in demonstrating the inadequacy both of the detractors – who predict numerous catastrophes provoked by technological innovations – and of the highly enthusiastic, who, even today, defend the idea that, thanks to the *new media*, knowledge will no longer be the privilege of a few chosen people, since the increased possibility of making thought a public issue will assure its extraordinary production. Leaving aside these two positions, there emerges, nevertheless, an empirical truth, whose negative effects risk being noticed only after a dangerous delay: the success of the Internet is implicit in the fact that it does not possess any intrinsic motivating or ennobling capability in the same way as, for

1 The concept of *user* that has come into force with computing innovation makes reference to the concept of *use* (from the Latin *utor*: I make use of), and it is strongly connected to services. The *users*, therefore, are those who use something, and whose particular essence consists in their having a pragmatic orientation. The *user* has also entered traditional fields, such as that of public communication, and those of institutions, wherein he or she seems to be the equivalent of the *client* of a public institution or domain. However, the term *client*, in a broad sense, maintains its standard meaning of 'purchaser of a service or product provided by a private entity' (see the introduction to this volume by M. Negrotti).

example, a trip to an unknown place – just as, on the Internet, everything depends on the 'eyes' of those who view it.

In fact, despite new technologies that allow information to circulate much more easily than in the past, we cannot affirm that this has determined a consequent rise in critical abilities, which, at the same time, implies that we surf the Internet, in the majority of cases, without knowing exactly what we are seeking. It is a kind of 'intelligent' entertainment which, counting on a little patience and an inexplicable *a priori* trust, sooner or later will reveal what we do not yet know. It will make us 'sail' towards previously unknown harbours, resolving, very often, the curiosities that might appear.

The creators of search-engines are well aware of this tendentious means of approach, to the point at which 'Google', for example, even has the pretension of interpreting the user's possible aim. Actually, by clicking on the key 'I'm Feeling Lucky', the search-engine projects the navigator towards the destination chosen by the same algorithm in an arbitrary and absolutely out-of-control way.

2 What future for art?

The aim of this work consists in trying to understand what effects the use of the Net can produce in the already fragile relationship between art and public. In general terms, what we are most interested in is an understanding of what evolution (if we can talk about such a thing) is implicit in passing from the enjoyment and knowledge of artistic facts via direct observation, to an indirect observation through the Net. In short, could art on the Internet induce, or on the contrary, reduce, the demand of undertaking a process leading to a mature aesthetic evaluation of the world and of the works of mankind?

The answer is certainly not easy to give, not least because, apart from the Internet, the nowadays privileged politics linked to the assessment of art shows that the adoption of facilitating strategies – for instance, the free entrance to museums – carried out precisely in order to bring the public closer to art is not necessarily effective, but in many cases, permits the appearance of pragmatic objectives of doubtful cultural consistency.

If we reflect, then, upon the repercussions that the mass media seem to have on all this, we might easily deduce that, notwithstanding the fact that successful exhibitions are often crowded – almost as happened a few decades ago with the Sunday morning mass – in reality, they do not take into account, more than superficially, the great importance of art or, even worse, of the cultural function of the artist.

The cultural 'flexibility' of our times permits us to 'dilute' the concept of art into a thousand little streams, and at the same time feeds us the illusion that, even taking it down to the most boring experiences of everyday life, its meaning will not disperse, and that, on the contrary, daily experiences will be ennobled. Certainly, at the base of these considerations, there remains the question, perhaps destined to remain ambiguously unresolved, of its ability to promote itself to any person, or rather, only to one who can claim a genuine propensity towards it. It is true that art addresses everybody, but not in the same way as does the air we breathe. It is, rather, something that needs a certain durability of reflection, in order to be digested and to become part of our patrimony of knowledge. For example, it makes it possible that the instrumental dismemberment of works, their use for commercial purposes, and their collocation in contexts that make their enjoyment more feasible (and precisely because of this, less penetrating), could represent an evident form of indulgence towards the demands and limitations of a 'user' whose interest is repeatedly orientated, implicitly and spontaneously, towards other objectives and values.

In order to keep our research from appearing as a kind of manifesto instigating a dogmatic 'technological witch-hunt', we wished to examine the design choices and contents of the web-pages through which the institutions – in particular, Italian town halls – transmit all of the information related to artistic events, especially Visual Arts, on the Net[2]. Obviously, institutions can assume a role that should not be ne-

2 The expression 'Visual Arts' applies to a very wide field which includes, apart from painting and sculpture, other artistic expressions such as video art, photography, installations and performance art. This broad sense of the term is the one given by the most important Italian institution for contemporary art: the Biennale

glected. If their communicative choices are dictated by common places or by superficial enthusiasm, the Net will be able only to computerize its inefficiency. If, by contrast, the institutions renounce political or technological short-cuts, they could assume a significant role in contributing to the deciphering of the intellectual significance present in artistic contents and manifestations.

For public institutions, the design of a web-site inevitably represents a strategic step, not so much because the result to be achieved is the production of an object in itself, but because there is the intention to establish a system in which the user and his or her learning become an integral part. For this reason, it is important, on the one hand, to simplify usage, even though this is not the only criterion to be considered. On the other hand, the face to face simplicity, openly recommended by *web usability* engineers[3], can become a double-edged sword: one who designs a public web-site will certainly have the aim of facilitating interaction, but will not necessarily think that all other considerations are subordinate. In other words, the risk we see round the corner, which must surely be avoided, is that the Web might reproduce exactly the same mistakes as those made by the traditional media. In short, if artistic knowledge becomes distorted or reduced when dispersing it, the Web could only offer, in the best case, a surrogate for aesthetic pleasure. Nobody denies that this can also be satisfying; but it must be remembered that T.V., radio, cinema and the Internet strike by means of broadcasting, taking advantage of the

---

di Venezia, which is inclined to distinguish these forms of expression from dance, cinema, theatre and architecture.

3 In designing web-sites, it is most important to take into account the user's part in accessing the site, and to seek the highest simplicity of the formal structure, of the navigational structure and of the organization and lay out of the contents. Sobriety in the design usually results in faster accessing of pages, and so is to be recommended up to a point, but without resulting in an excessive shortage of images. With regard to the wording, it is essential to employ clear and simple language, in order that it may be readily understood by as many people as possible. J. Nielsen has coined the acronym *HOME*, which synthesizes the four main qualities typical of a good website. It is formed from the expressions: *High-Quality*, meaning high-quality contents of pages; *Often*, meaning frequent updating; *Minimal*, meaning minimal times for downloading; and *Easy*, meaning ease of use [Nielsen, 2000, p. 380].

uncontrollable urges of the individual, and reducing most of us to merely passive users – 'telespectators', in the sense of those who watch from a distance.

3 An inquiry about the Web

The web-sites were analyzed as follows. We worked with a sample of 1350 Italian town halls, taken from a total of 8101 (ISTAT source), in order to form an idea of what constitutes the 'imagined' user of art for whom the various web-sites have been created. The sample obtained is of a probabilistic-cum-systematic type (at step 6), which is therefore, capable of providing, from a methodological perspective, a greater guarantee of representativeness and heterogeneity than that obtained through other forms of sampling.

Of the 1350 town halls analyzed, 31.2% (421) have an official web-site; 34.4% (465) have an unofficial site; and 34.4% (464) do not have a site. Obviously, at this point, our reference sample is composed of 886 town halls – namely those with web-sites. In the period June-October 2003, a pilot study was carried out with this sample, which was followed by a second observation during the same period a year later.

The analysis of the collected data, either during the first phase or during the second, then covered, essentially, two macro-areas:

1. The qualitative evaluation of the web-pages of every single town hall (considering: accessibility to the site, legibility, updating, and presence of cultural and artistic activities);
2. The statistic-descriptive evaluation of the whole distribution, through the employment of an electronic spreadsheet specially adapted for sociological inquiries (SPSS).

The analysis of the data focused on two main themes:

a. The use of the Internet by Public Administration;
b. The characteristics of the information provided.

As far as point A is concerned, it is noticeable that, in accordance with the trend of the last decade, the Internet is considered to be one of the communicative instruments adopted by the Public Administration (PA), but its use, as was predictable for such a young medium, seems more to be induced by the need of the PAs to demonstrate their being '*à la page*', rather than being conceived as a suitable means of communication.

Actually, only 48.5% of the official sites and 35.7% of the unofficial ones were updated in the last month. Such information leads us immediately to think that for the PAs it is relatively easy to create a new web-site, but difficult and costly to manage the site efficiently over time. The suspicion that computing technology could be a means to reduce the presence of manpower is not only mistaken, but is actually the opposite what is demonstrated by the data we have collected. In order to manage virtual information properly, a PA needs powerful machinery which unifies the technical competencies of the Data-Processing Centre, the Public Relations Office, the Town Hall Press Office, and the Mayor's Staff Office.

As we know, the PAs are caught between the increasing expectations of the public, on the one hand, and decreasing resources, on the other. While we can see that citizens are becoming ever more aware of their rights, with growing demands not only for a greater number of services, but also for higher standards, we can also see a fiscal crisis that takes away economical resources from public administration bodies at all levels, beginning with those of the town halls. We cannot avoid the fact that a reduction in resources will usually have a negative impact on the services provided. If, then, we consider that, in most cases, we a seeing an increment in the unevenness between external expertise and that of the PAs – which obviously suffer the consequences of both the impoverishment in economical resources, and the bureaucratic standardization from which there is no turning back – our premises appear to be more than justified.

This is once again proved by the fact that, with regard to our universe of 8101 town halls, we cannot talk about a true computerized model of the PAs, as only 47.5% of our universe of web-sites of Italian town halls make use of the standard *www.town-hall.name-of-town-hall.region.it* web-site address. The remaining 52.5% manage, in some way, to make do with posting a few web-pages on a wave of enthusiasm, often short-lived, which can only be to the detriment of the quality and usability of information provided.

However, this confirms that the institutions face numerous obstacles in the construction of their own web-sites, beginning with the fact that there are few people who possess the technical and didactic expertise necessary for its implementation. The lack of internal resources, in fact, leads institutions to become victims of self-styled 'experts', with potentially serious consequences.

It is interesting to underline, for example, that if we observe the data concerning the updating of official and unofficial sites in relation to the density of population, we can see that the differences relating to the updating process do not seem to be linked to the official or unofficial nature of the site, but to the number of inhabitants. In general, the more densely populated communities are the ones that give greater prominence to information on the Web. A circular, dated March 13th, 2001, provides guidelines relating to the organization, usability and accessibility of public administration web-sites, and claims, among other things, that 'the *Web* is a communicative instrument capable of modifying the flow of information within an organization, and between it and the outside world'. It also stresses that the 'prerequisite of any web-site design is the identification of the typologies of those to whom the site is addressed', and that therefore, 'it is important that an effort be made to imagine how the web-site will be seen and used by its visitors'. Yet in spite of the release of this circular, save for the information regarding the territorial surface of the community, the coat of arms, and just a few other pieces of information, in many cases it was not possible to recover even the indications relating to the composition of the assembly, nor, as a consequence, the effective competencies of the culture assessor.

It almost goes without saying that if, in general terms and through our observation, the web-site *home page* still presents itself as a fragile 'virtual business card' of the town hall, the information concerning culture can only be penalized further.

In fact, as far as we know, of the town-hall web-sites we examined, only 13.5% include a section addressing cultural issues, and an even smaller percentage, 6.7%, had content regarding Visual Arts. The term

Art without relating to the Visual Arts, was present in 7.1% of the cases.

A deeper study, which only corroborates the already cited data, must be carried out on the information concerning the composition of the town-hall assemblies, because it is precisely by paying attention to their presence, or lack of it, as well as to the role of the department of culture in each of them, that we can perceive that the destiny of art on the Web is already compromised, and that, given the premises supposedly, things will presumably not change in the future.

For example, if we consider the mere 137 official sites, together with the even more scanty 71 unofficial ones, in which the cultural department is represented, it might be interesting to see, for instance, what other skills are associated with this specific delegation. Actually, in 67.8% of the assemblies wherein the cultural delegation is present, we can observe that numerous other administrative fields are added to it. Notwithstanding the fact that this choice is understandable for some reasons, since the concentration of additional appointments is often dictated by contingent reasons, such as the small number of councillors in the smaller town halls, it is necessary to remark that it is not strange that culture is simply a reflection of the councillors' specific skills rather than the fruit of rational decisions made by the whole administration. Moreover, it should be emphasised that the presence in the municipality of cultural resources, such as the civic library and public or private art galleries, makes the cultural councillor the person in charge of managing these structures[4].

4 The combinations are of a heterogeneous and changing nature: ranging from the pairing *culture and urbanism* to the decidedly incongruous mixture of *culture and financial politics*. While a 'culture-orientated' approach seems to emerge in the various phases of administrative activity, one cannot deny the difficulty, if not the total impossibility, of putting together in the same department such contradictory demands as, for instance, the forced reduction of the library staff (after the implementation of unpopular measures, in particular case the increase in financial pressure) with plans to increase library lending. However, if we exclude the Urban Recovery Plans which, at least in their objectives, are inclined towards a revaluation of the historically determined building fabric, it is absolutely untenable to allow vehicular traffic in historical city centres whilst presuming to give them the title of 'City of Art'. We are dealing with ineliminable contradictions that are ever-present in public administration bodies, but which, at the same time, can assume,

If the specific capacities guide the assembly in choosing the specific charges of council departments, the often heterogeneous and changing associations can be rather interesting. Just for curiosity, and without claim of significance, we may mention curious aggregations between culture and particular delegations, such as 'cemetery services' with 'educational services, sports, free time and town-twinning' (town hall of San Mauro Torinese – TO); 'abusive waste and pit cleaning' with ' library services, promotional activity and images of Guarante, youth consultancy, adequacy of regulations, feasibility, street maintenance of every kind, and maintenance of public lighting and signposting' (Guarante town hall – CN); and 'energetics and technological installations' with 'staff issues, public instruction, scholastic buildings, computing, sports, and tourism' (Capracotta town hall – IS).

It is interesting to observe that, with regard to what we might call classical associations, the combination culture and sport, although still widespread, is beginning to lose ground (20%) in favour of combinations such as culture and scholastic instruction (29.1%) and culture and tourism (20%). The association culture and free time (7.9%) is an unequivocal sign of a changing society. Reflecting upon the percentages, it seems that, in contrast to the 1980s, when round-tables were organized to discuss how to spend free time, a new conception is developing which considers time as a precious resource. Concerning the pairing of culture and tourism, rather, we may perceive the leading role that the artistic patrimony imprints on the dynamics of attraction and receptivity of the territory. One is likely to think that the PAs have realized, rather late in the day, that, on the one hand, the artistic aspects, present to a greater or lesser extent in all Italian town halls, can represent a superb vehicle for the promotion of tourism, and that, on the other hand, we have sensed the informative possibilities that the Web offers for the widespread diffusion of these aspects. In fact, the presence of a web-site allows even those town halls excluded from the classical artistic circuits to enter into the market of artistic excursions, and to appeal, therefore, to the interests of an ever increasing propor-

to say the least, grotesque characteristics when these conflicts coexist after grouping together the delegations administered by the same councillor.

tion of tourists that are probably beginning to feel a little tired of the '*all-inclusive*' offers proposed by the agencies.

The analysis of the data suggests a reflection worthy of attention on the female presence among the town hall executives and on the meaning that such a presence can assume given the premises that we have previously discussed. If, on the one hand, the female presence in public affairs is to be understood as a significant sign of a maturity, late in coming, of the process towards gender equality, on the other hand, we must remark that the roles attributed to women are still rather marginal (this is perfectly consistent with the scarce female presence in the Italian Parliament), and this is in spite of the fact that of all institutional organisms, the town halls are those with the greatest proportion of women (17.9%) (according to data obtained from the Administrative Elections of 1997).

We shall not dwell on the possible causes that may determine such a scarce female presence, and that inevitably interact with other socio-cultural dynamics. We need mention only that even from our sample it emerges that there are typically male council departments and typically female ones, which in a certain sense would confirm a not altogether optimistic interpretation.

According to the Commission for Equal Opportunities, 2003, women are to be granted assignments related to welfare politics (30%), social politics (38.4%), culture (28.4%), healthcare (21.1%), and education (33.2%), whereas other assignments remain essentially male, such as those concerning public works (93.1%), urbanism and construction (89.9%) and the budget (91.6%). This suggests that gender differences are still particularly influential in the politics of the PAs: in short, the presence of women in public roles is accepted, but on the condition that this presence must not restrict the actions of departments that have always been considered most important.

In the specific case of our research, in the council departments of culture we found an average of 47.6% female participation; but this singular datum is potentially deceptive, especially when compared with the official datum provided by the Commission for Equal Opportunities, which gives a figure of 29.4%. Nevertheless, we must keep in mind the fact that this study examines only the information published on the town-hall web-sites, taking as reference, therefore, the data provided by those administrations that use the Web as a means of communication to the public. Thus, these percentages have been calculated – and it could not be otherwise – over the total of the town halls that indicated the presence of a councillor for culture within their assembly (23.5%). We are dealing, then, with a value that is intuitively far from the reality which nevertheless confirms the inaccuracy with which computerized information is managed, even those that should be most accurate and evident.

By contrast, the direct election of the mayor (Law N. 81, March 25th, 1993) clearly redefined a role which until that moment involved little autonomy, being 'compressed and compromised' between the pressures of the main parties and the instability of the executives. However, if the legislator's principal aim was the social participation and representation at all levels, we have no choice but to accept that Italian town halls remain inclined towards the employment of men, even with regard to the principal institutional figure. Moreover, the recent study by the Census (April 2003) on women in public administration bodies confirms an extremely significant datum: there are 635 mayors of communities with a population greater than 15,000 inhabitants, of whom only 46 are women; whereas in the smallest communities, out of a total of 7319 mayors, only 527 are women.

The low presence of women in the key roles of institutions leads us to think that, beyond the objective difficulties of participation and of the phenomenon of self-exclusion, the road towards the modernization of the PAs is still a long one, and only a genuine pluralism, beginning with gender pluralism, can guarantee the necessary representativeness of every request.

If we then observe the distribution of female councillors, 55% are in the north of Italy, 31% in the central region, and 14% in the south. Even though at first glance it would seem that women are recognized as having a higher sensibility towards artistic issues, our optimism breaks down in the face of the fact that, with regard to the distribution according to population density, the female participation reaches its peak, of 35.4%, within the communities with under 5000 inhabitants. This presence then drops to 27% in the communities with between 5000 and 10,000 inhabitants, before increasing slightly, to 28.3%, in those that reach 30,000 citizens; and finally almost disappearing completely (3.8%) in the communities with populations between 30,001 and 50,000 inhabitants, and those with a population over 50,000 (3.4% in regions with a number of inhabitants between 50,001 and 100,000, and 2.1% in those over 100,000).

We may conclude that in the cities of art and in big cities, where, presumably, there is no shortage of public or private financial resources, we should marvel at finding a woman councillor. Furthermore, if we add that the presence of the department of culture does not necessarily imply an increased focus on art – since the frequency of events does not correspond to the presence or absence of such a department – we can conclude that the PAs do not, presumably, attribute significant contractual power to such departments in the management of the available financial resources. Therefore, the female presence remains, so to speak, innocuous, or in any case irrelevant.

We now move on to point B, concerning the quality of art-related information on the Web, and in particular the frequency distribution, within our sample, of certain forms of expression belonging to the 'Visual Arts'. As is to be expected, given the premises, displays based on the art of painting, especially classical painting, represent a substantial proportion of the events. This shows that the PAs tend towards a limited knowledge of the world of contemporary art and of its more recent expressions. All is subordinated to the precept of not taking risks, and of therefore preferring those forms of expression that are 'scarcely contestable'. In fact, apart from the collectives of unknown local painters, which generally seem to adhere to the more traditional figurative forms, even the most consistent investment is far from being a perspective that reflects contemporary artistic work. Moreover, confirming that the town halls prefer to remain traditional, painting exhibitions are those that are able to attract the best sponsorships, even if in only 0.9% of the analyzed cases.

Very rarely, when an exhibition is announced, is all the necessary information provided to induce the citizen in possession of a minimal cultural interest to visit it. In fact, we never find the simultaneous presentation of information regarding the artists or artists – that is, whether it is an exhibition of the works of one artist or of many – the organizers, the typology of the works, and the dates and times of the exhibition. Yet these are indispensable pieces of information, needed to define an artistic event properly. Otherwise we are walking on the ground of the art of amateurs, with a thousand solutions or inventions which, in any case, have nothing to do with either art, in the strictest sense, or the correctness of the communicated information required.

What emerges in this pilot evaluation – awaiting further confirmation or refutation by a new and already planned observation in the future – is that on the studied web-sites the artistic events are essentially of two types: exhibitions of ancient art or art of the past, and exhibitions of local artists and amateurs.

This discovery leads us to the conclusion that the PAs do not turn to the official circuit of contemporary art when updating their web-sites, thereby enabling the public to obtain correct information. This, in turn, implies a further consequence; namely that not only does contemporary art in Italy lack a public image, but the Net, at least up till now, does not seem to be capable of overcoming this lack. It is as if the institutions, and consequently the public, notice neither their existence nor the cultural necessity to render public their forms of expression. In this sense, the figure of the user that results from the tested sites appears absolutely generic and undemanding: a mere *target* of a superficial institutional communication.

Our research, in short, proves that the information published by the PAs on the Internet does not help to foster an increment in the knowledge of contemporary art among the public at large, but leaves this duty – in the specific case of Italy – to the private circuit of galleries. It would be particularly relevant to introduce a curious hint regarding the difference that we can sense if we think of science on the Internet. G. O. Longo, in an article entitled 'Che cosa dice l'informazione?' [What does information say?] emphasises that the possibility of publishing one's own intuitions or research studies on the Web is creating a frantic acceleration of scientific work, which obstructs a well-considered deliberation on the results.

In the case of visual arts, it seems that just the opposite is happening; but the result, from a cultural point of view, is just as questionable: the information is so scarce that it can neither incite forms of authentic curiosity, nor push the public towards a deeper knowledge on this or that artist, nor induce anyone to invest part of his or her time to visit the exhibition. On the one hand, if the institutions do not misinterpret the kind of relationship that the public appears generally to prefer when interacting with new technologies, on the other hand, they also contribute to the making of art an object of *usage* rather than of knowledge.

4 Conclusions

From this initial investigation, we can derive some general considerations that show problematic conditions for a genuine attention to culture and, inevitably, to art:

1. There is a disconcerting scarcity of attention paid, on the part of public institutions, to the politics of cultural marketing, which should demonstrate a greater sensitivity towards the existing public cultural patrimony, as well as to the contemporary artistic culture present in official circuits, in Italy and elsewhere;
2. The experiences of the 1980s, of completely substituting the public cultural budget with private money, led to a policy of limiting the flow of resources almost exclusively to the most proven values; that is, those that would allow a greater visibility of the investment and a consequent increase in prestige. The result was an exodus from the circuit of cultural tourism of thousands of events, asphyxiated by lack of resources;
3. Despite the CENSIS study of 2002[5], which maintained that culture must 'give sense' (or lead, or offer a guide), art seems, for many institutions, to be a fundamentally unproductive investment, almost irrelevant to the 'true' needs of mankind.

Our research is, in a certain sense, a confirmation that the cultural facts, and, even more, those linked to the visual arts, are not generally perceived by the PAs in terms of social responsibility. In other words, it appears clearly that even today there is not a *politics of and for culture*.

Yet that is what also emerges from a study regarding the propensity to adopt new technologies, with respect to the use that is made of them, and which could not but have had an influence on art on the

5 We refer to the study 'I giovani e la cultura nell'era della comunicazione' ('Young Persons and Culture in the Communication Era').

Web. We refer to the *Report on innovation and digital technologies* promoted by the Ministry of Innovation and Technology, attended by M. Morcellini, wherein it is pointed out how successful it is considered to appeal to the ludic rather than to the functional dimension [cf. Morcellini, 2003].

As far as art is concerned, we should add finally that it is certainly not a case either of censuring the artist for his or her enigmatic or incomprehensible work, or of accepting, uncritically, that the public does not seem to search for tools to understand that work. It is inevitable, though, that we should reflect upon the fact that the detachment between the world of art and the public is producing a series of cultural fragilities in both, and we certainly cannot say that the PAs are truly aware of it. On the one hand, this detachment has perceptibly distanced the thoughts of the artist from the customary contents of existence, reducing them to a self-referential system, which at times seems to have become an artistic system. On the other hand, concerning the public, we assist in the disappearance of the requests for clarifications and perceptive stimuli that only artists can offer – an attitude that is in itself disputable, and which inevitably recalls the conclusions reached by Le Bon and by Ortega y Gasset, in relation to the loss of interest of the masses in all that which is non-conformist: that is, to the different, the singular, the qualified and the original.

Bibliography

K. S. Gill (ed.), *Human Machine Symbiosis: The Foundations of Human-Centred Systems Design*, Springer, London 1996.

M. Morcellini, 'Società dell'informazione – Gli italiani alla prova del nuovo. Rivoluzioni nella comunicazione e nuovi comportamenti culturali', in <http://www.techrev.it/technologyrewiev/index.php?p=article&a=126>, 2003.

J. Nielsen, *Web Usability*, Apogeo, Milano 2000.

Danila Bertasio, LCA – Laboratory for the Culture of the Artificial, University of Urbino

Elena Bettinelli

Religions on the Internet: a comparison between different kinds of websites and users

Abstract:

Religions and new media represent an innovative and often fascinating union.

Both the three main monotheistic religions and the other forms of cult appear to need to be represented on the net and on other hypermedia and multimedia circles

The approach is peculiar to each religion and, according to each doctrinal and historical substratum, all religions create a communicative model that enables the individual to relate with the rest of the community: for instance, the Catholic Church presents itself through a coherent and organised web structure and places the 'user-believer' at the very centre of the informational network.

The Islamic approach tends instead to establish, in digital codes, a route to emphasise the role and the potentials of the 'ummah' (the community of the believers) thus giving priority to an ideal which is specific to that community.

Geographical and medial territory also represent an important means of mediation: religions tend to create an equilibrium most suitable to their dogmatic structure balancing their geographical location (seen as the place of origin of the faith and endowed with symbolic significance) and the medial territory as space of new proselytism.

Keywords:
Religions, individual, community, Internet, website, user, believer, faith, monotheistic, sectarian, Christianity, Islam, Catholic, physical territory, territory of media.

1

The meeting between the current networks of communication and the need of experimenting new ways of spiritual and religious research describes a phenomena that only has contradictory features at a first glance.

The proceeding to insert the religious message into the context of the hypermedia technology suggests an important attempt to mediate two issues of only formally contradictory nature.

On one hand the strained relations with the religious leaves behind long debates that have, depending on the case, emphasized or reduced his/her anthropologic value. The religion itself represents a very old system: the reflections that concern the religion have gone through some continuous changes, corrections and second thoughts throughout time.

On the other hand, the system of the so called 'new medium' reveals, starting from the name itself, an unstable and at the same time precarious balance always in need of new definitions: the instruments of the communication hardly manage to maintain their promises and, chasing an up-to-dateness, they finish being in some way out-of-date when they eventually become technically realized.

We talk about the religion as an ancient spirit, incapable of ageing, and about Internet as a so recent spirit that it paradoxically reveals itself to be 'old' and in the research of a continuous updating they find themselves having a necessity in common, destined to merge itself and to produce results that reveal themselves to be at the least curious.

2

It is first of all important to make a distinction between the different groups of users who derive from at least the same amount of different needs of affiliations as the religions.

Even if Christianity and Islam seem to have considerable differences at the doctrinarian and dogmatic level and so also from the point of view of the value, they both have similar relationship as far as the faith is concerned: both religions share a vocation of a universalistic type. And they therefore address the entire mankind.

Hebraism, which is one of the three monotheistic 'sisters', is to be found on another level. While it is fundamentally based on an ethnic base it does not imply the necessity to establish a faith based on vocation: the principle that concerns the proselytism does not therefore concern the Jewish community that applies a relatively simple principal of recognition of the individual in a vision of community. The Jewish foundation is based on ties of blood, on the family: The Jew is a Jew because of a particular condition linked to his/her birth of being the child of a Jewish woman and/or be willing to accept the *halacha*, the normative corpus. The *halacha is* not only civil nor purely religious, but with an aspect that involve all the relationships that connects the single individual, the community and God.

To be Jewish is not either restricted by the religious practices: even if a Jew lives as an atheist he maintains the status of being a Jew. A reciprocally inverted process occurs with the Muslim who lives with the practices of Islam with the meaning, 'submission to God', where the necessary qualification, that is connected to his/her faith, is an individual and deliberate act of conjunction with God.

The dissemination on Internet refers itself to preserve the traditional remembrance, the entire spiritual, normative and artistic heritage that has distinguished Hebraism. Many web sites become the place where the religious establishes his/her historical identity. The websites address in particular an international audience with the solicitation of moral condemnation and refusal for what represented the anti-Semitism and anti-Zionism. The geopolitical legal actions in the Middle East territory, the conflictive situation between Israel and Palestine are elaborated through a historical background from which a complex array of attributed rights and self proclamations emerge. Israel represents the Hebrew people, their spiritual, religious and political[1] identity: The roots of Hebraism are founded upon an ethnic-religious belief that is extremely attached to the territory and its story is concentrated on a particular area, a land that fixes and circumscribes a unique identity, but from which also cultural values of an national and universal context emerge.

The gathering of characteristics mentioned above reveals something anomalous: as far as the communication is concerned Hebraism does not show any significant urge to disseminate the religion through

1 Cf. Castelnuovo, 2001.

missionary, unlike what occurs in the Christian and Muslim religions. The hope to obtain conversion or at least to be able to enchant the person enough to persuade him/her to establish a first contact represents an important investment.

The religious enthusiasm towards the conquest of the means of the communication and to obtain the total control as far as their codes and specialized languages are concerned is characterized by different choices, but all of them, that are at the same time rivals, dedicate themselves to construct an ethic convincing project with the help of the technology and its users.

3

Internet has some extraordinary features. It is decentralized, interactive, characterized by immediacy, used all over the world and also expandable in an indeterminable way for when it comes to the contents. It is also flexible, very adaptable and egalitarian, with the meaning that whoever has the necessary instruments and a modest technical knowledge can actively be present in the cyberspace, transmit his/her message and search for an audience. It consents anonymity, role-play and the opportunity for the user loose herself/himself in day-dreams in the environment of a community. Depending on the taste of the user it can contemporary be an active participation as a passive absorption in a world 'of narcissistic and egoistic incentives'.

The directives according to which *the approach of the Catholic Church* organizes its interest concerning social communication and Internet, considered as *marvellous technical inventions (...) 'instruments of God' destined, according to the Providence, to unite mankind in brotherly ties, in order to make them collaborators of His plan of Salvation* are to be found in the books *La Chiesa e Internet* (Church and Internet) and *Etica in Internet* (Ethics on Internet).

The Catholicism sets an informative system with the aim to reproduce, in the language that are suited to the new mean, what the ecclesiastical magistery has elaborated. The Vatican website and the presence of websites that have been organized as portals *(Portale*

Cattolico, Trinitas, Siti cattolici, Chiesa cattolica, Totustuus, Tantumergo, Profeta, Netcrim ...) organize the contents in a independent net with the purpose to create a sort of strict, authoritative top-down subnet to which a wide range of informative resources are linked. These resources, more or less loosely managed cover the existing area in between the formal ecclesiastical magistery, that is to be found at the top, and the activities connected to the catholic ethics that are developed in a concrete and direct reality: missionary, ecumenical and spiritual gathering activities addressed to the young population and to the families. These persons find a reference on the Internet as much as in the structures that are more sensible to the necessity of a physical territory as the parish, orders, congregations etc.

The 'circuit of Catholic websites' develops through a redundant structure where the user several times is invited to get in touch with different types of websites, that are named 'verified', 'the most visited ones', 'the most voted ones', 'the most clicked upon', 'recommended'. This makes it possible to monitor the users.

This approach is a great opportunity: to be able to listen and to dialogue. It is technically realized through 'electronic grids', correspondence with enclosed nuns, telematic presbyters – a service offered by the website *preti on line* (priests online) –, retreat online, prayers of intercession, links of poetic or stimulating inspiration that give access to areas where to publish a prayer, a message, a dedication ...

As far as the Christian-catholic structure is concerned the current relationship between 'the individual' and 'the community' has a preference for a centrifugal force; starting from the community in search of the individual. This becomes an approach that tries to reach all the persons involved in the process of religious communication. The tendency of affiliation that models the believer's belonging to the community emphasizes the role and the value of the single human being. Thus a net of verifications is created for the religious who justify them as useful. A project of cooptation emerges therefore with the intent to strengthen the initial similarity as far as the religious structure from the beginning is concerned. That kind of similarity, even though it is existing, is not strong and needs a more mature identification[2].

According to this hypothesis, the user appears as a believer-user, thought of as a single element, which gets to know, even though only superficially, the basic thoughts of the Christian evangelization and therefore considered a 'natural' member of the community of faith. The believer assumes a role in the 'low level of abstraction'[3]: not as a 'person' in search of a collocation based on value and identity but above all as a 'believer' deriving from a gathering of specific doctrinarian knowledge and not willing to be confused with mixed and syncretic spiritual identities.

The proliferation of websites that call themselves Catholic creates a different kind of problem. As we have already mentioned, the groups connected with the church should be present in a creative way in Internet.

*Also the non official, very motivated and well-informed individuals and groups that act on their own initiatives have likewise the right to be present. It is however a motive of confusion, at the very least, not to distinguish eccentric doctrinal interpretations, bizarre devotions and ideological proclamations that are called 'Catholic' from the authentic positions of the church (*taken from *Chiesa e Internet).*

The intention to expel the elements that are potentially capable of confounding the doctrinal pureness of the Catholic legitimacy, is justified by the necessity to preserve a doctrinal and normative accumulation of a century-old authority. The Catholic religious communication expects however that these fundaments should be in some way adapted, 'reprogrammed' according to a specialized language adapted to the technology of the Internet. The church must be at its believer-users disposal.

2 For an overview of the religious situation in Italy see Garelli, 1996; De Vita and Berti, 2001; Garelli, Guizzardi, Pace, 2003.

3 The current environment of research considers the user as independent and places him/her in a middle position within a hypothetic scale of abstraction. Its highest value is represented by terms (as person, individual) deriving from disciplines that use extensive and general descriptions. The lowest values concern the terminologies capable of identifying contexts and specific functions of application (client, spectator, contributor ...).

The Catholic church finds itself being involved in an Evangelic upgrading[4], above all in favour of using the means of the social communication in a new, modern and flexible way and among which the Internet stands out for its celebrity and capacity. Its users are to be found on different levels and progressively 'made into believers' according to a top-down system, from the formally reliable community to the single human being.

4

The balance between the importance of the single human being and the value of the community is established through opposite forces in the Muslim environment, based on the community, compared to the Catholic religion, based on the single human being. The Muslim believer founds his/her identity when giving a concrete form to an imaginary, multiform, not compact Islamic community, with the expectation to find a theoretical structure that is able to create a more coherent and stable identity.

If the Net may serve as a formative and educational place of a superior level compared to a primary religious education with less assimilation for the Christians, Internet becomes an occasion of public visualising for Islam where to merge a process of comparison, self description and challenge, within sight of a strengthening of the identity.

Muslims Online in an effort dedicated to promote the use of the Internet within the Muslim Community. With the explosive growth of the Internet being a fairly recent phenomena, we feel it is imperative for Muslims to generate and produce their own content on the World Wide Web. Inshallah, this site will develop into a central location where Muslims can read content, exchange ideas, and keep abreast of community events. Alhumdulillah, the Internet, like the Muslim world has no bounds, and we hope that this new medium will help in uniting and strengthening our ummah by widening the channels of communication between our Muslim brothers and sisters (taken from the website Muslims Online).

4 Cf. Casanova, 2000.

From this emerges the desire of a clear representation, with the intention to re-establish, the *ummah* – the community of the believers – in the globalized era[5], giving it a concrete aspect to expose in a virtual debate where it could be possible to reflect and to become aware of the self perception.

The telematic Islam[6] virtually 'pursues' the Muslim only to be able to reflect him/her into the community. Islam wants to create a base of public agreement where each person represents the engine of a centripetal cohesion with the aim to eventually realize a united vision. It all therefore deals with an extensive, horizontal, bottom-up recovery of an opposite kind compared to the one hypothesized for the Catholic organisation.

The Islamic resources, unlike the Catholic ones, can not count on a homogeneous substratum, on a severely defined orthodoxy to transfer into the world of relations of the Internet. Islam is transferred in a cultural ideal that has to be realized. It is not limited of a Dogmatic structure which is subject to shared religious codes. It is instead a matter of a greater, all-embracing ethic ideal extensively spread out over the society where the variables involved are numerous, complex and mutually related[7].

The gathered content in the Islamic websites collocate spiritual and ecumenical approaches, basic doctrinarian norms, the pillars of the faith, history and art with the notions and proclamations of the political-ideological faith from the more or less strong virulence, aspects and debates that we could call of a civil problematic nature, of scientific knowledge, a cultural as well as religious coexistence, to the inclusion of compromises with atheist conditions that fluctuate between technology and marketing. Its aspect is characterized by complexity and contradiction provided with diversified and occasionally interactive characteristics. The tension that derives from a disorganized ensemble of currents, law schools, the complicated mechanisms that are meant to absorb the modernity[8], the coexistence of cultural pro-

5 Cf. Kurtz, 2000.

6 Cf. Bunt, 2000 and 2003.

7 Cf. Allam, 2002.

8 Cf. Mernissi, 2002.

grammes, the conflictive antagonism that distinguish the geopolitical panorama[9], the presence of fundamentalisms with a distant tribal ritualising that bursts inside the domestic television screens 'nearby': all these phenomena contribute to a vague idea of how demanding it must be for Islam to circumscribe a negotiable cultural reach as well as when it comes to the faith and also a historical, anthropological position in the globalized world that is dominated by the communicative arteries of people's consent and mistrust.

5

Not only the religious aspect of monotheistic inspiration opts for an established appearance within the telematic and hypermedia dimension, but also a wide and diversified range of beliefs and spiritual currents deriving from the most varied origins prove to have found their own space too.

The typical syncretic coexistence[10] of many and different traditions, cosmogonies, pantheisms, conceptions of the nature and what transcends the empiric existence, local divinities, astrology, herbal medicine, amulets is fully realized in the websites.

The structure of the net seems to adapt itself very well to a less strict kind of religious and philosophic reading compared to the more well-established structures of faith. Every religious reading becomes, just as the hypertextual one, something capable of recomposing numerous tracks of interpretative ideas and senses of personal belief. Just as the reader on the net is active, with the aim to construct a personal collection of references and contents, the user of the 'esoteric and mystic nebula' is backed when it comes to the necessity of creating a personal architecture of his/her proper need of the faith.

The suggested contents and links reveal a variety as wide as it is incoherent of topics related to a holistic conception of the individual. One of the countless pieces of the puzzle that forms the world and the universe.

9 Cf. Fuller and Lesser, 1996
10 Cf. Lucà Trombetta, 2003.

The more these beliefs diverge from a well-structured pattern of religious doctrine, the less they are forced to compete against each other, discrediting one another. Faiths of this kind are free from the necessity to put themselves in contradiction with a different and pre-existing religious consciousness. This fusion of doctrines, new age movements and of 'human potentials', Oriental practices and philosophy, which loose their original identity in the Western countries, favours the education where individual is able to do the most undiscriminating choices. The criteria of inspiration are however often emphasize the efficiency. The single believer is persuaded to live according to a holistic way of life, that includes body and mind, but through the most productive management of his/her knowledge of psychological resources.

Horoscopes, tarots, herbal cosmetics, *Feng Shui*, acupuncture, graphology, alternative psychotherapy, fortune-telling, angles, stones, talismans, positive thinking are all associated with other more traditional religions in the search engines for the spiritual topics. They associate for example with Taoism, Shivaism, Animism, Buddhism and not to neglect with occultism, parapsychology, ufology, cults of fays and elves, 'the left hand path', Satanism. A more than suitable metaphor to describe the confused but determined spiritual sense of the informative, technological era that we live in where the web is the witness.

The user that relies on such a flexible spirituality, on a melting-pot of believes and convictions, is to be found on the opposite side compared to the Christian and Islamic religious experiences in the Internet. Even though Islam and Christianity propose different patterns of belonging based on the individual-community, they both maintain a call for something original, an irreplaceable epicentre from where their way of thinking derive in the world of faith mediated *online*.

What differs them from the sectarian-philosophical setting is that the proposal of faith for the last mentioned does not have to conform coherently with something already given, something already certain. This corresponds pretty well with the mere ambition of an individual free from specific rules of belonging, to a user that could be specified as detached from his/her territory as regards to precise references of value: even the partly joining to a certain monotheistic faith – that is jealous by definition –, in short a betrayal, does not involve serious responsibilities. The 'believer' of this kind is rather inspired by a vague religious idea capable of regaining and adapting in a useful way whatever conviction that allows a free, mutable, flexible choice. He/She becomes therefore more disposed to the 'inversion' than to the 'transformation', to sudden changes of the route: it is almost like a metaphor of the hypertextual reading in a religious vein.

In a global society[11] that theorizes the right to obtain the spiritual wellbeing like whatever informative element, the links gives access and exploitation with just a 'click' to 'religious products'. Not only the ways of how to connect are plenty but also the user himself/herself is able to structure a path of faith *ad hoc*, home-made, custom-made gaining a lot in a prospect of the concept of net[12], but loosing himself/herself in the multiplicity of connections. It deals however with the most manageable part as far as the potential of Internet as a whole is concerned: the polytheistic sectarianism forms a divided user in the middle of numerous combinations and informative/religious contaminations. It also assists the user to cut, at least cognitively, his/her bounds with the community and the physical territory. He/she is an individual in a full and absolute sense, located in a geographical environment that does not describe the 'physical territory' but the 'territory of the media', an isolated element in the middle of different structures of communitarian and ecclesiastic belonging.

But it all deals with a concept that, even though it promises – and partially actually maintains – freedom of consciousness and choice, it makes it paradoxically a burden, loading it with a heavy 'burden of choices'[13] where the weight is destined to increase and to multiply the load of the appearing options. The believer is therefore, increased to a 'high level of abstraction', in every aspect a religious user of an already mentioned 'risk society'[14].

11 Cf. Barman, 2001.
12 Cf. Castells, 2002.
13 Cf. Berger, 1994 and 1995.
14 Cf. Beck, 2000.

Bibliography

K. F. Allam, *L'Islam globale*, Rizzoli, Milano 2002.

Z. Bauman, *Dentro la globalizzazione. Le conseguenze sulle persone*, Laterza, Roma-Bari 2001.

U. Beck, *Che cos'è la globalizzazione. Rischi e prospettive della società planetaria*, Carocci, Roma 1999.

U. Beck, *La società del rischio. Verso una seconda modernità*, Carocci, Roma 2000.

P. L. Berger, *Una gloria remota. Avere fede nell'epoca del pluralismo*, il Mulino, Bologna 1994.

P. L. Berger, *Il brusio degli angeli. Il sacro nella società contemporanea*, il Mulino, Bologna 1995.

G. R. Bunt, *Virtually Islamic. Computer-mediated Communication and Cyber Islamic Environments*, University of Wales, Cardiff 2000.

G. R. Bunt, *Islam in the Digital Age. E-Jihad, Online Fatwas and Cyber Islamic Environments*, Pluto Press, London 2003.

J. Casanova, *Oltre la secolarizzazione. Le religioni alla riconquista della sfera pubblica*, il Mulino, Bologna 2000.

M. Castells, *La nascita della società in Rete*, EGEA, Milano 2002.

A. Castelnuovo, 'Ebraismo e pluralismo nell'età moderna', in R. De Vita e F. Berti (a cura di), *La religione nella società dell'incertezza. Per una convivenza solidale in una società multireligiosa*, FrancoAngeli, Milano 2001.

R. De Vita e F. Berti (a cura di), *La religione nella società dell'incertezza. Per una convivenza solidale in una società multireligiosa*, FrancoAngeli, Milano 2001.

Etica in Internet, Pontificio Consiglio delle Comunicazioni Sociali. <http://www.vatican.va/roman_curia/pontifical_councils/pccs/documents/rc_pc_pccs_doc_20020228_ethics-internet_it.html>

G. E. Fuller and I. O. Lesser, *Geopolitica dell'Islam*, Donzelli, Roma 1996.

F. Garelli, *Forza della religione e debolezza della fede*, il Mulino, Bologna 1996.

F. Garelli, G. Guizzardi, E. Pace, *Un singolare pluralismo. Indagine sul pluralismo morale e religioso degli italiani*, il Mulino, Bologna 2003.

L. R. Kurtz, *Le religioni nell'era della globalizzazione. Una prospettiva sociologica*, il Mulino, Bologna 2000.

La Chiesa e Internet, Pontificio Consiglio delle Comunicazioni Sociali. <http://www.vatican.va/roman_curia/pontifical_councils/pccs/documents/rc_pc_pccs_doc_20020228_church-internet_it.html>

P. Lucà Trombetta, 'Sincretismo e mercato religioso', in AA. VV., *Religioni e Società, Il fardello della ricchezza nelle etiche religiose*, Università degli Studi di Firenze, maggio-agosto 2003.

F. Mernissi, *Islam e democrazia. La paura della modernità*, Giunti, Firenze 2002.

Religioni e Società, Il fardello della ricchezza nelle etiche religiose, Università degli Studi di Firenze, maggio-agosto 2003.

Religioni e Società, La trascendenza e il destino dell'Occidente, Università degli Studi di Firenze, maggio-agosto 2004.
G. Romano, *Il punto sulla qualità dei siti di carattere religioso*, in atti del Convegno *Internet: un nuovo forum per proclamare il Vangelo*, Milano 2002.

Elena Bettinelli, Facoltà di Scienze Politiche, Università di Trieste

Monica Bordegoni, Umberto Cugini and Rossella Mana

Novel interaction methods exploit designers' skills

Abstract:

The paper describes novel interaction methods and technologies that have been recently developed for supporting the design and validation of products in the industrial design sector. They are based on the paradigm of virtual prototyping, make use of innovative techniques such as virtual, augmented and mixed reality, and innovative interaction devices, such as haptic devices, and aim at exploiting designers' skills.

Keywords:
product design, haptics, Mixed Reality, designers' skills

1 Introduction

The product design phase is subject to several recycles where design choices are reviewed several times on the basis of testing results. When confidence in product design is reached design is released for manufacturing. Mistakes in the design phase have the most impact, and the amount of subsequent rework downstream is still substantial. It is well known that the more design changes occur closer to the manufacturing phase, the more the impact on product development costs increases. In order to reduce mistakes and design changes, practices performing design testing and validation efficiently and effectively – starting since the early conceptual design phase – have developed.

A common and traditional practice makes use of physical prototypes for performing validation tests. Actually, the design phase is supported by CAS (Computer Aided Styling) and CAD (Computer

Aided Design) tools that produce and release digital descriptions of products. Therefore, mathematical models are immediately available for subsequent re-use. In the recent years, several tools supporting virtual validation and simulation of products have been developed. The spreading of these tools has been also supported by increasingly powerful hardware enabling the execution of complex simulations. The practice based on simulation is named 'virtual prototyping' (Virtual Concepts Conf., 2003).

Compared to physical prototyping practice, *virtual prototypes* are in general less expensive, allow for not one but several simulation runs on a single model, are easily configurable and support variants. Furthermore, tests are repeatable, and results of validation are often immediately available for product design review. We can say that virtual prototypes often provide insights that physical testing would not reveal. Anyway, if digital prototyping does not completely substitute yet physical tests, they help optimizing and eliminating redundancy in test facilities, accelerating life testing, etc.

Currently, companies tend to reduce the use and numbers of physical prototypes, opting for using them at the very end of the design phases, and extensively using digital prototyping early in product design (Fig. 1).

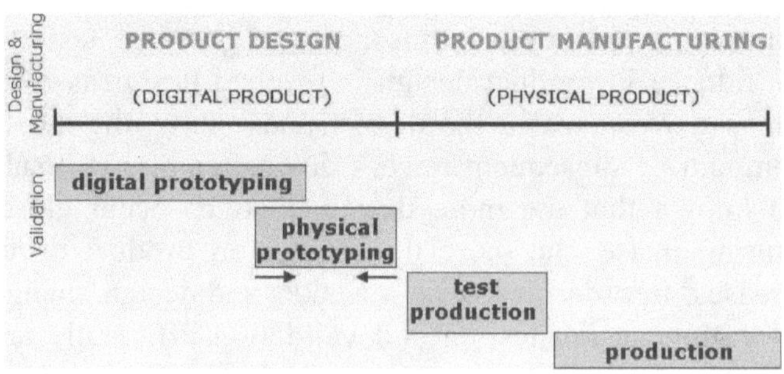

Fig. 1. Product design and manufacturing phases

Companies are interested in adopting methods and technologies for supporting design process in an effective and efficient way, in a more and more competitive context. That would allow the reduction of cycles, developing time, and therefore costs, while increasing product quality.

Traditional design, typical of mechanical sector (design of machines and of mechanical systems) mainly puts emphasis on those aspects such as product functionality, dimensioning, reliability, etc. Computer aided design tools (CAD systems) have been developed with the aim of satisfying these aspects, offering designers an environment that supports design activities in an effective way. The industrial design sector differs from the one of traditional design since the product development process has to answer to some requirements typical of the sector, such as: is the product nice? Is it easily usable by the final user? Does it satisfy some ergonomics requirements?, and so on. Tools aided to industrial design favor aspects like modeling and realistic visualization of product shape, global evaluation of the product (considering both context and environment), contextual evaluation of the product (considering the interaction of the user with the product itself). Therefore, techniques and tools aided industrial design are different from those developed for traditional design.

The paper describes novel interaction methods and technologies that have been recently developed for supporting the design and validation of products in the industrial design sector. They are based on the paradigm of virtual prototyping, make use of innovative techniques such as virtual, augmented and mixed reality, and innovative interaction devices, such as haptic devices, and aim at exploiting designers' skills.

2 Product Design

A typical practice of the industrial deign sector is the use of maquette and physical prototypes (physical mock-up – PMU) in the initial concept phase of products. Modelers are used to produce PMU of products by modeling and manipulating malleable materials like clay, foam material, wood, etc. Having in mind the target shape of product, modelers use their tacit knowledge they have in using hands for controlling the generation and modification of the shape. The skill they have developed in doing that is usually a valuable asset to be preserved for a company. Unfortunately, PMUs require time to be developed, are expensive and not easy to test. Today, this phase is followed by a Reverse Engineering phase where the geometric model is reconstructed by acquiring a data set describing the object shape using some 3D acquisition technology. The process of shape reconstruction is time consuming and error prone, since acquired data have to be aligned, merged together, holes have to be bridged, surface has to be decimated and smoothed, precision properly controlled, and so on. Once the final 3D model is obtained, it can be imported in a virtual prototyping environment, for realistically rendering it, for performing animations, simulations, etc. Fig. 2 shows an example of steps of an iron design process going from the initial maquette to the final virtual prototype.

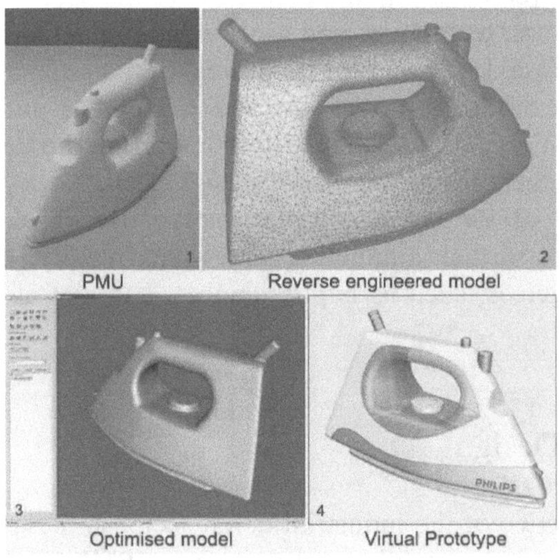

Fig. 2. Design process of an iron: from maquette to virtual prototype

These traditional practices are today used together with or replaced by practices that take advantage of the use of computers. CAS (Computer

Aided Styling) and CAID systems (Computer Aided Industrial Design) are today extensively used for virtual prototyping of products. Shapes and features (for example, materials, colors, behaviors) of products are described using appropriate techniques and tools that are used in an effective way both at the initial conceptual phase and at the subsequent engineering phase. Functionalities of CAS/CAD systems have improved offering designers support for capturing, reusing and sharing design elements. Besides, they also support checking shape precision, surface curvature (using reflection lines), etc. They are used in several industrial sectors (appliances, fashion design, automotive, etc.). Unlike PMU practice, virtual prototypes are less expensive, less costly and easier to test.

Despite of the advancement of design tools functionalities, complex shapes can be defined using advanced modeling operators, but anyway using a mouse or a pen to interact with. Especially in the industrial design sector, this modality does not exploit users' skills. In fact, modelers and designers are used to make products using their hands, as well as to feel the quality of the shape of the material they are molding by physically touching objects. Designers are dissatisfied using mouse & keyboard to design 3D shapes, and would like to be physically more active in the product creation phases. Therefore, CAD tools require to be enriched with more user friendly and effective ways of interaction, not only based on mouse & keyboard, but using more powerful and realistic interaction tools allowing for the exploitation of modelers' skills.

3 Product validation

Within product lifecycle process, design and validation plays an important role. Two kinds of validation of the design of a product are usually performed:

1. validation of functional aspects;
2. validation of ergonomics issues.

In the following, some examples of product validation are provides.

3.1 Validation of functional aspects

The validation of functional aspects aims at testing product behaviour and engineering aspects, like aerodynamics, crash, stress, deformation, noise, vibration, etc. Today, these tests can be done on the virtual prototype using analysis and simulation tools. These tools allow building sorts of virtual lab tests that are often comparable to physical hardware ones.

Particularly innovative are physics-based simulation techniques that simulate bodies' behaviour, applying the laws of physics (Barr, 1997). This technique is typically used in cloth and clothing simulation field in order to obtain 3D virtual prototypes that allow designers to get some useful information for the validation of aesthetic and functional aspects of the designed cloth. That allows the reduction of the number and role of physical prototypes required in the various phases of garment development process. (Fontana et. al, 2004). The physics-based technique allows designers to simulate different levels of fabric complexities: from simple cloth (Fig. 3) to complex-shaped fabric structure.

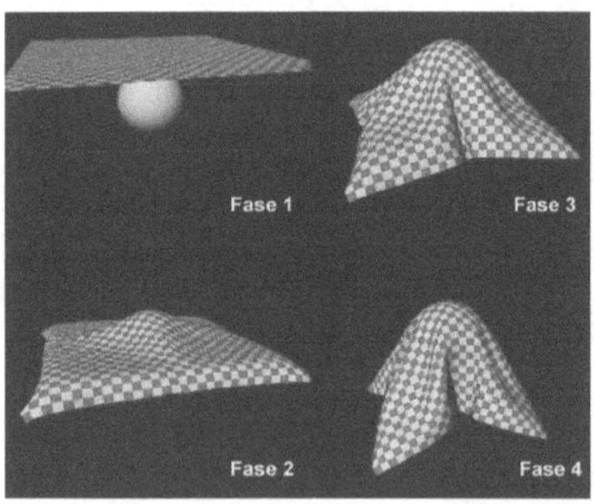

Fig. 3. Simulation steps of a piece of cloth draping over a sphere object

In order to simulate realistic cloth behaviour we have used a 3D garment simulator, named SoftWorld, developed by the KAEMaRT group (www.kaemart.it). This simulator adopts a particle-based model where thin deformable objects are described through a discrete filter of particles having mass properties, subjected to internal and external interactions (Cugini and Rizzi, 2002). As and example, we have simulated a man's jacket. This is a significant test case of a real-size labored garment. The virtual prototype of the jacket is obtained assembling a large number of 2D panels, corresponding to several components (e.g. front, back and side parts, sleeves, collar, lapel, etc.) positioned onto a 3D parametric mannequin, by means of sewing rules and other construction constraints. Moreover, it is important for an accurate simulation of real tailored garments to consider properties of the textile materials, and interactions with the external environment. Fig. 4 shows the multi-layered structure of a jacket front and some simulation results obtained varying the mechanical properties of the material for reproducing the ironing action.

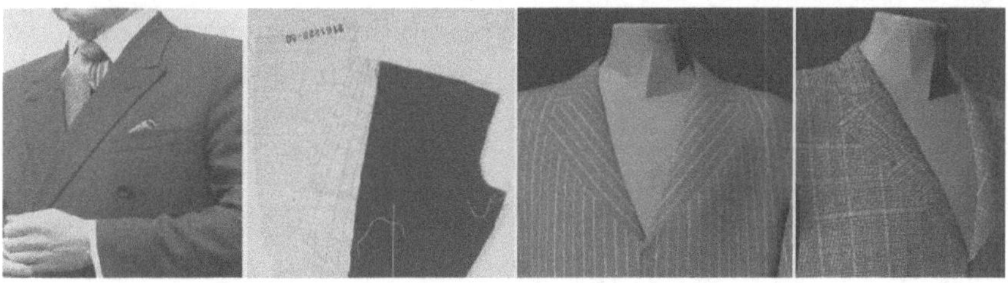

Fig. 4. Multy-layered structure of the jacket front and rever pleat simulation results

3.2 Validation of ergonomics aspects

For what concerns the validation of ergonomics issues, ergonomics standard points out three aspects to consider (Ergonomics, 2001):

- Effectiveness: if and how a task (objective) is achieved;
- Efficiency: effort required for achieving a pre-defined level of effectiveness;
- Satisfaction: degree of comfort perceived by the user.

Ergonomics can be described as the application of scientific information concerning humans to the design of objects, systems and environment for human use. Some types of ergonomics validation are run using virtual prototypes. For example, safety tests are run successfully in virtual environments. Virtual mannequins have been integrated into design tools in order to support a preliminary validation of products. A typical application is the use of virtual humans for validating the driver's field of view from inside the car (Culubert Montanera, et al. 2001).

Actually, the validation of several ergonomics aspects of products, like comfort, usability, efficiency, effectiveness, etc. are subject to personal feelings and tastes of users, related to their preferences, emotions, skill, cultural and social aspects, etc. Computer representations are not able today to fully address every ergonomics issue in the product validation phase; therefore physical prototypes and the intervention of real users may still prove necessary.

In the following, we analyze the possibility of using real/virtual users for running ergonomics tests of real/virtual products (Fig. 5).

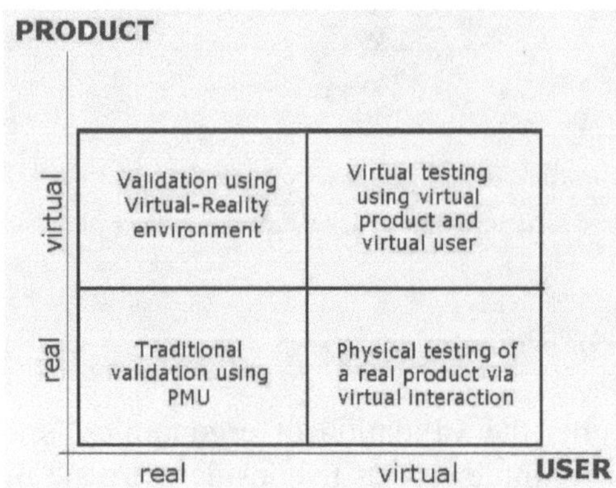

Fig. 5. Relation between real/virtual user and product concerning the product validation phase

Real user/Real product: Users perform traditional tests using physical prototypes. Most of the ergonomics aspects can be validated but physical prototypes are expensive to realize and not easily configurable.

Virtual user/Virtual product: Virtual humans are used for testing few aspects of products, through the simulation of digital models. Some tools support the definition of computational models of humans, the computation of art positioning, and therefore the prediction of postures and comfort. It is possible testing user comfort by measuring virtual human posture parameters like bending degree of limbs, forces exerted on the seat, etc. While several aspects can be tested in a virtual environment, conversely, there are not models that can be used for predicting manipulation forces applied by users.

Real user/Virtual product: users perform tests using virtual prototyping of products and environment. Current state of the art of Virtual Prototyping tools provides realistic visualization of objects. Conversely, the level of interaction is still poor. More powerful and realistic interaction modalities with virtual products are required, especially for the validation of ergonomics aspects (Di Girolamo et al., 2001).

Virtual user/Real product: Real products are tested by means of virtual interaction simulated by a machine.

The following section presents novel methods to enrich interaction with Virtual Prototypes.

4 Novel interaction method and technology: haptics

Interaction methods based on haptics are changing the way of interacting with virtual models, by adding tactile information to visual one (Bordegoni et. al, 2002). Some new devices appeared in the last years on the market, named haptic devices. The term 'haptic' is used to indicate both tactile and kinestetics aspects of the sense of touch. Some devices can be worn on a finger, like the PHANToM devices (SensAble Technologies); some others are exoskeleton reflecting force on the user's hand, like the Sarcos Dextrous Arm Master; there are actuated gloves, like the Virtual Technologies Cybergrasp (Virtual Technologies), and bi-directional active devices with three degrees of freedom, like the Haptic Master (FCS-CS), and several others. While the above mentioned devices are general-purpose, there are other devices that are oriented to specific applications, especially for the medical field. According to this new interaction modality, users are able to perceive objects characteristics by looking at them on the monitor, and by touching them by means of the haptic device. Fig. 6 shows one of the authors interacting with a virtual model of a product using a HapticMaster device.

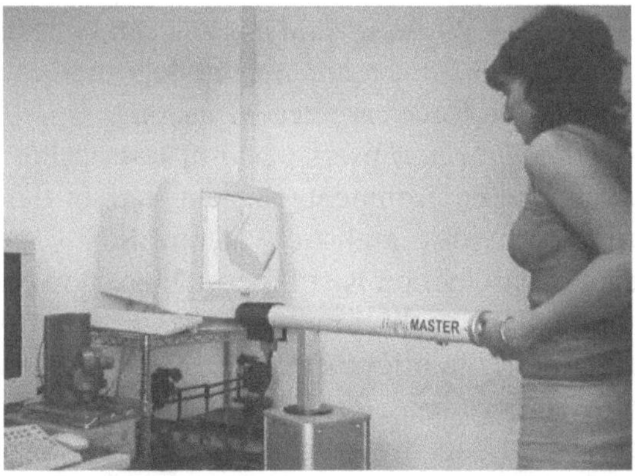

Fig. 6. User interacting with a Virtual Prototype of a product using a HapticMaster device.

4.1 Use of haptics in design

Haptic devices have been proposed to be used in product modeling. An example is FreeForm by SensAble (www.sensable.com) that is used for virtual sculpting. Users can see and feel the shape of the product they model, and can also have a feeling of the type of material they are moulding (clay, foam material, hard material, etc.). This system simulates a spherical sculptor's tool used to remove or add material from/onto an initial block of material. The limit of this system is that the device is point-based, and therefore constraints the type of interaction to be through tools like pens or sticks, but not full hands. Furthermore, high aesthetic quality level required by several industrial design sectors (like, the car industry or high-end customer products) is not reached since the spherical tools provided by the system are not sufficient to generate complex shapes. This technique is anyway promising and it would be interesting combining this innovative user interface with a more aesthetic intent driven process.

Currently, the KAEMaRT Group at Politecnico di Milano is coordinating a research project financially supported by the FP6 IST Programme of the European Union named *T'n D – Touch and Design* (www.kaemart.it/touch-and-design) that aims at developing a novel system for shape generation and modification based on novel haptic interaction and intelligent shape manipulation operators in order to exploit manual skill of modelers and designers. Designers will be able to work in a similar manner as modelers adopt in their daily work, using a novel haptic device (and extension of FCS-HapticMaster), for modeling product shapes. Haptic technology will be integrated together with modeling operators for providing a working environment offering high degree of usability and acceptance, comfort of usage, flexibility for any kind of user, including inexperienced users, that would support the exploitation of users' skills.

4.2 Use of haptics in validation

The validation of ergonomics aspects of a product includes the validation of its 'look and feel'. Both these aspects require the involvement of real users in the validation process. Users interact with the representation – real or virtual – of products in order to test its 'look' (attractiveness, colour, textures, etc.) and 'feel' (behaviour, textures, etc.). While the first involves only aspects related to vision, the latter involves aspects related to physical contact with objects. This is typical of several products, especially consumer products (like control devices of car dashboards and airplane cockpits, washing machines, ovens, music/video players, mobile phones, etc.), but also industrial products (like control devices of machine tools).

The use of physical prototypes in ergonomics validation has several disadvantages and limitations. In particular, variants and configurations cannot be quickly set up and tested. Therefore, it seems interesting studying the interaction of real humans and virtual products.

In order to understand which are the capabilities and effectiveness of ergonomics validation performed using virtual prototyping, it is necessary to have a view of the state-of-the-art of technologies enabling virtual prototyping. Systems where users interact with virtual environments make use of *Virtual Reality* techniques and technologies (Earnshaw, 1995) (Fan Dai, 1988). Mainly, Virtual Reality state-of-the-art consists of visualization techniques like Head Mounted Display – HMD (Virtual Research Systems), large screens, workbench, CAVE, and others, that give users the impression of being immersed into the virtual world. Besides, more realism is provided by stereo-viewing technology. All mentioned technologies allow the perception of visual aspects (look) of products.

On the other hand, those tasks involving interaction of users with systems require more appropriate media and modalities. Some technologies have appeared recently and support innovative ways of interaction: whole hand devices (Virtual Technologies), 3D devices (Logitech), etc. The major limitation of these devices concerning tasks like validating interactive devices (ergonomics testing) is the fact that they do not provide information about the sense of touch.

State-of-the-art of general-purpose haptic devices supports the simulation of activities like 'inspect an object', 'push a button'. Other more complex tasks, like 'turn a knob' are difficult, if not impossible, to perform with these devices. This is mainly due to the current development stage of commercial devices that do not allow the simulation of complex situations like 6 d.o.f. operations, operations requiring torque computation, rendering of high forces, and so on.

Alternative approaches for those kinds of validation consist of *configurable hardware prototypes*. Most of the examples have been realised in the automotive sector. Some research centers of automotive companies use physical prototypes that allow for the configuration of several types of vehicles, by modifying some parameters, like position, orientation, etc., of some physical parts. These solutions are usually very expensive and require long re-configuration time, and are mainly developed for a unique product type.

An alternative approach to the use of haptic devices and configurable hardware prototypes consists of dynamic haptic systems whose behavior can be programmed. Dynamic haptic systems consist of ad hoc devices that can be programmed and configured for simulating the behavior of specific objects. For demonstrating the concept, at KAEMaRT Group researchers have designed and built some haptic systems (for example, a dynamic knob prototype) that can be dynamically programmed for the validation of ergonomics and usability design aspects. It consists of a mechanical system, a controller connected to a computer running a program for setting the knob parameters in real-time. The knob prototype can be mounted and positioned into an environment. For example, it can be mounted on a car dashboard or on a white goods panel board. Users can be asked to try and use the knob. If they do not like its response and behavior, and want to modify it, the knob parameters can be changed in real-time. When the preferred configuration is reached, corresponding parameters and data can be stored by the control system for subsequent use in design review phases.

These hardware prototypes simulate products behavior, but usually do not implement aspects concerning the 'look' of a product. Usually, they are ugly mechanical systems, simulating the behavior of a device, but not the appearance of both the device and the whole environment. For example, the knob developed can be used to simulate a knob mounted on a car dashboard. In order to provide users a view of how the whole environment will be once the car is built, knob visual aspects are not satisfactory as they are implemented in the hardware prototype: the visual appearance of the knob does not look real and cannot be changed; besides that, the users do not see the environment where the knob should be mounted on (the car dashboard in our example). On the other hand, if we provide the visual representation of the car dashboards only, the user will not be able to validate the ergonomics aspects of devices mounted on the dashboard, like the knob. A solution for overcoming this problem is mixing real and virtual information.

5 Mixed real-virtual prototyping

When only real or only virtual information is available in a system, it might not be enough by itself so as users can fully interact with the system and accomplish some given tasks. In some cases, it might be useful conceiving systems that mix virtual and real information.

These systems are made of two parts (Fig. 7):

- one part consists of real environment and objects;
- the other part consists of virtual representation of both environment and objects.

An application can implement the human-computer interaction as a mix of virtual and real modalities, where virtual and real information is present at variable percentage. If the information fully comes from the real world we talk about *Reality*; if the information fully comes from the virtual world we talk about *Virtuality*. If both the modalities are present in the application, we talk about *Mixed Real-Virtual* interaction modality.

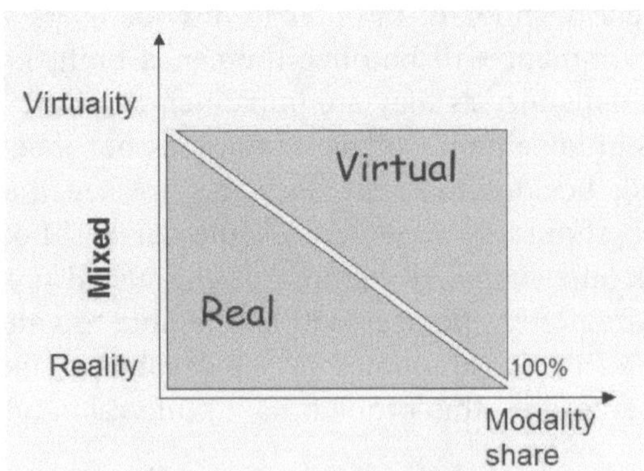

Fig. 7. Mixing real and virtual information, according to different modality share

From a practical point of view, mixed real-virtual interaction can be implemented adding information to the already existing one. If a system is primarily based on real information, the quantity of information that can be made available for use can be improved adding virtual information to it. These systems are called *Augmented Reality* systems (Milgram, 1994). The quantity of information available in the augmented interaction interface is increased of a value that depends on the quantity of virtual information that is added.

Conversely, if an application is mainly based on virtual representation of environment and objects, real objects can be added to them. In this case, we talk about *Augmented Virtuality* systems. Also in this case, the quantity of information available in the augmented interaction interface is increased of a value that depends on the quantity of real information that is added. The ergonomics application using the dynamic haptic system we have developed is of this type. In fact, a realistic visual representation of the environment and the interaction devices is usually available, since it comes from the design phase (CAD models). We have added information about those characteristics of the knob that cannot be perceived through the virtual information only, by integrating a real physical device into the system. The virtual information (digital model of car dashboard and knob) is augmented with real information (physical dynamic systems) yielding an Augmented Virtuality system.

6 Conclusions

This paper has presented the results of a research work we have carried out for experimenting novel methods to perform design and validation of industrial design products, aiming at exploiting designers' skills. We have shown examples of the use of the innovative physics-based approach for realizing virtual prototypes of cloth and clothing. This application is particularly useful for garment stylists in the initial conceptual phase of garment design, where the possibility to produce virtual prototypes of garments enable the evaluation of several solutions. Besides, we have shown some examples where novel haptic technology is used for the validation of ergonomics aspects of interactive devices. The experiment demonstrated that virtual prototyping, if augmented with the sense of touch, can be successfully used for the validation of ergonomics aspects, that are critical to evaluate in a purely visual environment.

The experiments carried out so far show that recently developed technology allows designers to use applications and tools in a more user friendly and intuitive way, where tactile senses, besides vision, can be used in the design and validation phase. More experiments are going to be set up, for improving the method and proposing it for other application areas.

Bibliography

A. H. Barr, A. Witkin, *Introduction to Physically-Based Modeling,* Course Notes, ACM SIGGRAPH, 1997.

M. Bordegoni, U. Cugini, 'Visual-Haptic Multimodal Human Computer Interaction', in M. Negrotti (ed.), *Yearbook of the Artificial – Methodological Aspects and Cultural Issues*, Vol. 1, Peter Lang, Bern, Berlin 2002.

U. Cugini, C. Rizzi, *3D Design and simulation of men garments*, Proc. Workshop WSCG '02, 2002.

C. Montanera, G. Tovar Gomis, 'Use of a Virtual Human for ergonomics analysis of a car cockpit', *Laurea Thesis UPC – Barcelona and Dip. Meccanica – Politecnico di Milano* (Erasmus programme), 2002.

G. Di Girolamo, G. Monacelli, M. Martorelli, G. Vaudo, *Use of Virtual Mock-ups for Ergonomics Design*, Proc. of 7th International Conference ATA, Florence 2001.

Ergonomics. International Encyclopaedia of Ergonomics and Human Factors, Vol. 1, Taylor and Francis, London 2001.

R. A. Earnshaw, M. A. Gigante, H. J. Jones, *Virtual Reality Systems*, Academic Press 1995.

FCS-CS, HapticMaster, <http://www.fcs-cs.com/robotics/robotics2.htm>.

M. Fontana, R. Mana, C. Rizzi, U. Cugini, 'Capturing Shape and Complexity of Tailored Apparel within a 3D CAD Framework', *2004 INDUSTRIAL SIMULATION CONFERENCE (ISC'04)*, June 7-9, 2004 – University of Malaga, Malaga 2004, pp. 523-528.

Logitech, <http://www.logitech.com>.

P. Milgram, H. Takemura, A. Utsumi, F. Kishino, 'Augmented Reality: a class of displays on the reality-virtuality continuum', *SPIE Vol. 2351*, Telemanipulator and Telepresence technologies, 1994.

MMVR 2003, Proceedings of Medicine Meets Virtual Reality Conference 2003, Newport Beach, CA, 2003.

SensAble Technologies Inc., <http://www.sensable.com>.

Virtual Concepts Conference, Proceedings, Biarritz, 5-7 November 2003.
Virtual Reality for Industrial Applications, Fan Dai (Ed.), Springer 1998.
Virtual Technologies Inc., <http://www.virtex.com>.

M. Bordegoni, U. Cugini and R. Mana, KAEMaRT Group, Dipartimento di Meccanica, Politecnico di Milano

Luciano Celi

'It's like if': the power of exhibits in Science Centres and Museums between immersion and illusion

Abstract:

A short survey shows how the artificial – in a broader sense – is used at present in museums and science centers. The latter in particular have upset the identity of museums often changing their aim: from the ancient mottos like 'look but do not touch' or 'do not touch' to – through the experience of Frank Oppenheimer with the Exploratorium *in San Francisco (USA) – 'it is forbidden not to touch', with interesting experiences in Europe – in Great Britain with Richard Gregory and the* Exploratory, *recently merged in* @Bristol *– and in Italy. In this transition, exhibits have had a fundamental role: from very simple objects, like a Möbius tape with verses written upon it that can be re-read in a different way by cutting the tape, to very complex and involving objects like performances and interactions in virtual reality. All according to an old pedagogic maxim which says that it is useful to 'learn while having fun', which in English is resumed in a single word:* edutainment.

Keywords:
Science Centre, Museum, exhibit, virtual reality, informal learning

1 History

As an introduction, it is necessary to say that this is not the right context for tracing a history of museums – especially of science museums – and that it would be very difficult to tell such a history, since museums have (had) several foundational ideas and intricate histories, varying from place to place, making the field we are trying to move in more heterogeneous and complicated. It seems therefore useful to provide some background elements to the history of museums here, being it quite a recent one, we are now attempting to describe.

1.1 From Ulisse Aldrovandi's 'theatre of the world' (1522-1655) to the 18th century

Science museums have their origin in the 16th century as collections of natural objects, accomplishing a more and more increasing demand:

- the intellectual need of 'tidying up' the confused taxonomic hierarchies generated by the introduction of new species, above all vegetal species, as a consequence of the discovery of the American continent;
- the disappearance of a proper classification system[1] that, leaving behind the middle-age model, be able to offer something more than a simple juxtaposition with no criteria other than the morphological one, inherited from Aristotle.

Similarly, it can be considered as the transition from the middle age pictorial representation to the perspective age: it is no longer possible to place everything on the same level as in the precedent herbariums and bestiaries. It becomes then necessary to interpret nature and divine project and, in this sense, Aldrovandi[2] is among the initiators of a tradition that sees the accumulation of new knowledge; he creates rooms located according to a precise criterion (a location common to Imperato's[3] visions of nature) for which minerals are placed on the low-

1 We remember that Linneo (1707-1780) will comes after in this history.

2 His motto was 'nullus ordo, nulla cognitio': without order it's impossible having cognition and comprehension. We can find this *weltanschauung* in more different fields: the art of French and Italian gardens is a good example, because it's a way – maybe the best – to make order in the Nature. In the same way this will happens when the German 'kammer' (religious or not) was created: the aim consists in the representation of the king power's and often the power was proportional to the complexity, dimensions and richness of the 'kammer'.

3 Ferrante Imperato (1550-1631) made a naturalistic museum to Palazzo Gravina, in Naples (his home). In brief his museum becomes one of the most famous in Europe and was visited from a large number of researchers. By this contacts and his frequent travels in South-Italy, Imperato increased his collection and made

est level, vegetal species on a higher level, animals on an upper level, up to fish and birds, which are usually placed at ceiling level.

Table 1: *Visitors to Ulisse Aldrovandi's Museum Until 1605*[4]

Visitors	Number	Visitors	Number
Popes	1	Professors	21
Legates and Vice-Legates	7	Philosophers	6
Papal Governors	4	University Graduates	21
Cardinals	10	Scholars	907[5]
Archbishops	11	Doctors	87
Bishops	21	Anatomists	1
Theologians	15	Surgeons	1
Inquisitors	3	Apothecaries	6
Other Clergy	190	Distillers	1
Princes	6	Mathematicians	2
Nobles	118	Cosmographers	1
Courtiers	4	Antiquarians	1
Diplomats and Princely Servants	7	Printers	2
Famous Men	26	Booksellers	2
Appointed and Elected Officials	26	Painters	6
Lawyers	43	Poets	1
Secretaries	8	Jewellers	1
Notaries	1	Women	1[6]
Military Leaders	6	Unknown	4

good geological observations that confirm the water's role to modelling coasts and mountains.

4 Source: BUB, *Aldrovandi,* ms. 110 *(Catalogus virorum, qui visitarunt Musaeum nostrum, et manu propria subscripserun in nostris libris Musaei. Secundum ordinem dignitatum, studiorum, et professionum)*. Aldrovandi and his assistants recorded 1579 visitors to the museum.

5 The category *studiosus* included one entry of indeterminate number, that we have discounted from the tally.

6 The only woman visitor listed was Ippolita Paleotti, listed as *studiosa.*

The aim of the curators is to make the syntax of the world explicit; the aim of visitors is – be they academics or dukes – to know how and where to place objects: the mnemo-technic exercise, apparently useless, has the advantage of making the underlying structure of nature more evident, divine mirror.

The word 'museum'[7] has, actually, ancient origins: it comes from the ancient Greek *museion,* bearing the precise meaning of 'temple dear to muses' and, in that sense, there existed temples in Alexandria in Egypt in the third century B. C., which hosted religious and academic people. In the 16th century the word is used again because its origin was associated with a place of study, with a precise taxonomic and didactical value. Later, or in a way parallel, came the words *Kunst* and *Wunderkammer* that, even though not belonging to the Italian tradition, are often identified as the origin of the modern museum. The main difference between the Italian 'teatri' and the German 'Kammer' is in their aims: in Italy there is an attempt to interpret and to seek for the divine order through nature[8], whereas in the German Kammer the objective is to cause astonishment by means of fantastic objects.

1.2 On wonder: 'naturalia' *and* 'artificialia', *prodigious and miraculous*

The Kammer tradition has ecclesiastic origins and comes directly from the churches' *thesaurus*: to owe fantastic and rare objects is a sign both of material possession and of intellectual and almost magical domination of the world: the attribution of rareness, wonder and preciousness of the object were immediately transferred onto its owner and typically the owners were the Church or, alternatively princes or powerful men who could afford to collect various kinds of objects.

7 That, in this period, was used as alternative to 'theatre'.

8 Often with objects that nothing have to do with the 'marvellous': the aim is not surprising the visitors because in Italy we arrive from S. Agostino's philosophy, where all phenomenon are manifestation of the divinity.

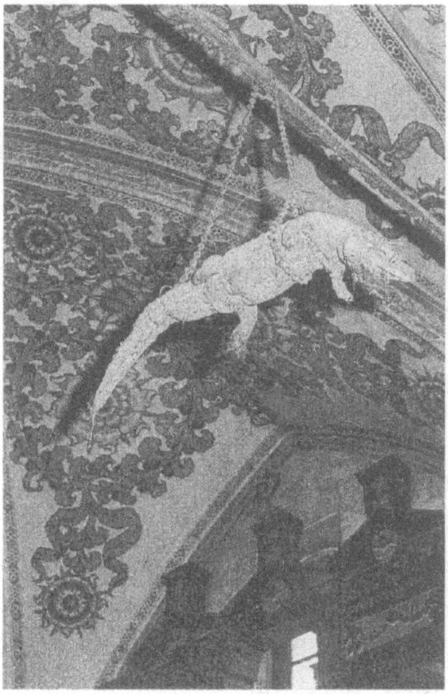

Fig. 1. Interior of the Church Santa Maria delle Grazie in Mantova[9]

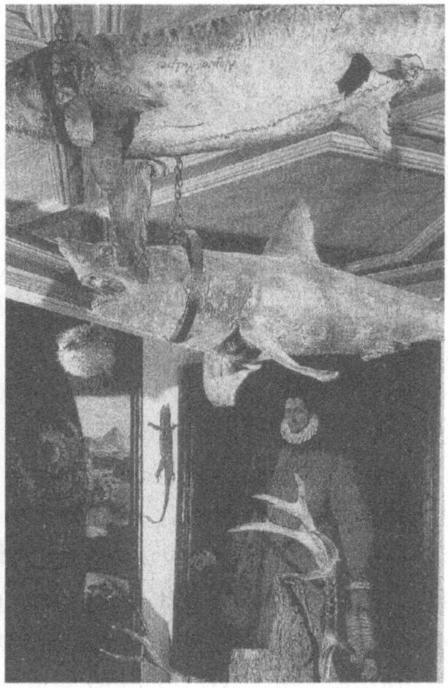

Fig. 2. Interior of the Ambras castle in Innsbruck

All this comes from a 'pan-marvellous' tradition: everything contemplated by mankind, from the most common phenomena, like the rain falling, to the most inexplicable, like sun eclipses, constitutes wonder, even though with some distinctions: the 'naturalia' belong to a higher status because they are connected to the divine sphere; the ancient Greek 'φυσιχη' means nature, but have the same root of 'φυο' that in his intransitive form means 'to born, to rise, to sprout' and Aristotle himself right in his *Physics* has a very high consideration of nature because of this feature. 'Artificialia', made by mankind, no matter how prodigious they can be, are kept in a lesser consideration: they do not become and have therefore a minor strength[10]; their ability to become

9 Both figures are in Von Schlosser (2000).

10 The famous seven wonders of the ancient world are cited as an example of men's ability, but highlighting how they be subjected to time's corruption. Even Shakespeare, in his *Winter's tale*, treats as 'artificialia' and denigrates flowers grown in

is just that of becoming corrupt by the action of time. From the 13th century on, this vision is more and more modified: much of the Greek culture, arriving to us from the Arabs, offers a different sense from this distinction between artificial and natural, cutting this umbilical cord connecting God and nature.

A different distinction to the one just proposed concerns the several shades of marvellous: what is marvellous is placed at the world's borders or comes from the world's borders and this precise meaning, coincides with 'what I don't know because too far away'. Something prodigious and miraculous can happen in our reality, here and now and it is the kind of exception – the two-headed child – that can be contemplated because it is a divine manifestation.

In the 17th century, among the aims of museums there was the money-earning prosaic one, in order to keep alive and enrich collections. In this sense, the exposition to exceptional is aimed to – as it happens in many modern exhibitions[11] – the astonishment and to create a magic halo which also characterized Atanasius Kircher's[12] museum in Rome: from an architectural point of view, we find walking passages and wide galleries constituting a kind of 'catabasis', a kind of initiatory trip for the visitor, who can see astrological and propitiatory symbols in front of himself; to the physical walking, a metaphysi-

---

gardens and therefore less 'natural': Perdita, about to get married, says: 'Sir, the year growing ancient, / Not yet on summer's death, nor on the birth / Of trembling winter, the fairest flowers o' the season / Are our carnations and streak'd gillyvors, / Which some call nature's bastards: of that kind / Our rustic garden's barren; and I care not / To get slips of them.' And, to Polixene's request of why so much resentment, she replies: 'For I have heard it said. There is an *art* which in their piedness shares / With great creating nature.' (my italic for 'art').

11 For example, Gunther Von Hagens' *Körperwelten*.

12 Atanasius Kircher (1601-1680), Jesuit, in 1551 opens his collection to the visitors: his *Wunderkammer* was founded on Alfonso Donnini's donation, that has especially ancient objects, added with other important pieces collected from Kircher himself and his religious brother in America and China. All object are described in two great catalogues and the first of them was published from De Sepi in 1678. The collection is the greatest for quality and quantity of objects and it's possible to think it as the real forerunner of all naturalistic and ethnographic museums. In 1773 the Jesuit's Companion was suppressed from Clemente XIV: the consequence of this event is the first division and dispersion of kircherian's collection.

cal one immediately adds to it, being it necessary to benefit of the right star conjunctions[13]. The story of the Kircherian museum will be, *mutatis mutandis*, the same for the Wunderkammer which, with ups and downs, will leave some space for museums in their modern structure.

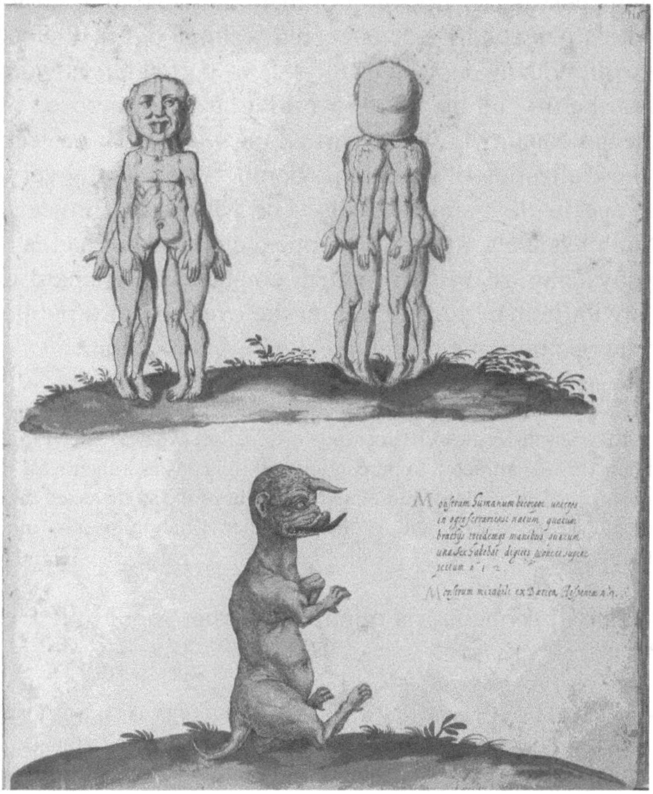

Fig. 3. *Monstrum humanum bicorpor* and *monstrum mirabile* from Aldrovandi's catalogue

13 It seems that also Tommaso Campanella, before dying, designed in his bedroom an astral conjunction to 'change' his destiny.

2 From history up to now: from museums to science centres

We are in the immediate post-war period, overseas, in the United States. The power of the atomic bombs fallen on Hiroshima and Nagasaki has made most of the world's public opinion silent. It is clear to the world the destructive potential of mankind, able to destroy himself and the whole planet earth; science and technology, that should be to his service show their dark side. Physics in particular, is won by this trial: Enrico Fermi and his whole working group composed by scientists and technicians, who were part of the Manhattan project, whose aim was the realization of the atomic bomb, have actually very little to celebrate, due to their ability/skills. The Manhattan project also involved the physician Robert J. Oppenheimer who had a brother, Frank, a physician as well. He got in contact with Richard Gregory, who deals with human perception, on the occasion of a public lecture just after the opening of a museum called Exploratorium:

> Shortly after its opening in 1969, I happened to give a lecture in San Francisco (the first Smith-Kettlewell Lecture, on illusion) where I met Frank Oppenheimer. He was attracted to perception and illusions as a theme for the Exploratorium (and very likely had already thought of it), so he asked me to design perceptual experiments, and to discuss in depth the philosophy of the whole enterprise.[14]

But what's in Oppenheimer's mind? And what should the Exploratorium be?

> The proposal that Oppenheimer brought [to San Francisco] was for a museum in which people would directly experience and manipulate things, instead of being told about them. The public was to interact with objects as an experimental scientist does in the natural world or in the laboratory. The museum was to teach that the subject matter of science is all around us and that its comprehension is available to all. It was to remove science from the exclusive

14 Gregory (2000), p. 1273.

> domain of experts, to demystify it, and to restore it to the common sphere. It was to convince that doing science can be interesting and fun for everyone.[15]

This is the reason why the exhibits[16] were born, becoming then the instrument through which to realize this revolutionary kind of museum. This museum will not be called museum any more but will become *Science Center*, or else a place where – opposite to what happens in traditional museums – the motto will be 'NOT touching is forbidden', according to what today has become a real category through which museums can be classified: *hands on*[17].

Oppenheimer's idea was with no doubt an original one and if its aim was – even though just partially – to attract people, especially the young generations, to science, it succeeded. The idea of Exploratorium was exported and 'adopted' in many places of both the old[18] and the new continents, where, at the same time, the Exploratorium was evolving, including scientists, artists and craftsmen who built new exhibits, improved and took care of the maintenance of the existing ones, often leaving the doors of the 'laboratory' open to the public, whereas initially it was an off-limits place.

15 Hein (1990), p. XV, quoted in Gregory (2000), p. 1273. Official history uses this date as birth of the first Science Centre, but the idea to use 'machines' (exhibits) where an interactivity with visitors is possible in museums, was present in Europe since 1925, with the Deutsches Museum.

16 Gregory (2000) also offers the genesis of the word: 'Although in the 14th century "exhibit" meant to *hold*, from the Latin *exhibitere*, and the French ex – our + habere = to hold, it came to mean to *display* on view', p. 1274 (footnote). In the history of museology we can find that objects initially used with didactic aims in the 18th century, are exhibits *in nuce*: more of anatomic reproductions – that comes from Italian's ceroplastic school – was used in Obstetrical schools, but when their main 'function' (instrument to learning the correct techniques to help in childbirth moment) stop, they were collected in museums (for example, la Specola, in Florence). See also Celi (2004).

17 Versus *hands off*. Often the museum – in more traditional way of think it – has a conservative finality and normally between objects and visitors there is a good glass…

18 Gregory himself, after the experience with Oppenheimer, ten years later will open in Great Britain the *Exploratory*, recently incorporated in *@Bristol*.

3 From simplicity to complexity, between illusion and immersion

The genesis of this alternative to the traditional museum has, as we could see, an impression linked to the interests of the two founding members, Gregory and Oppenheimer: perception.

At the beginning of the 20th century, the use of diorama started to get widespread in the United States: this was a background, as a sort of *trompe l'œil*, able to create an environment in which it was possible to contextualize the displayed objects and able to attract the visitor's attention. A lot has been said about this technique; recently its usefulness and its importance have been discussed, so much as to make 'illusion' turn into 'immersion'[19].

> A scientific exhibition essentially consists of real objects and phenomena. An exhibition is, then, 'concentrated reality' withdrawn from its natural environment and reorganized with intelligence and beauty. Part of a museologist's job, therefore, is to create new environments, more intensely in some areas than in others. We expose general aspects and specific solutions in the following areas: an Amazonian rain forest, a shipwreck, a collection of insects trapped in amber in the Oligocene epoch, a fall of meteorites, an infectious epidemic, thoughts on inert, live and cultured forms, and thoughts on the great revolution in physics.[20]

3.1 Illusions to learn with no risks: virtual reality of the Cybernarium

Virtual reality (VR) can be used with several aims: in the case of the Cybernarium, in Germany, its main aim is *edutainment*, that is learning through entertainment. In an interview with Stefan Müller, coordi-

19 Obviously there are important differences between them: with word 'illusion' we will intend an interaction that deceive our senses and our perceptions (in this sense virtual reality is an illusion); with 'immersion' we will intend our being physically in situations that seems to be real.

20 Jorge Wagensberg (2003), abstract of his talk, where he stressed on the importance to 'create the right atmosphere' and proposed an interesting comparison between natural ambient and Gaudi's art. On this argument see also Oppenheimer (1973).

nator of the internal research group, it is possible to find the aim of the Cybernarium and the objectives of the researches:

> In the future, human beings will need the expansion of their virtual world more and more and in the possible future, this technical approach will be more and more successful. In everyday life, we can create a direct approach with the virtual world through our home computer. Mankind needs an interactive relationship with machines and students and the school are our target. In schools there would be the possibility to widen the discourse on physics and biology, microcosm and macrocosm through this interaction dynamics We […] think that the main need and demand of people nowadays is to try and experiment. Our research group is made of 21 people who are carrying out a virtuality scientific project, with an attempt to re-create virtually a real world.

Besides,

> […] it is possible to see artistic details of a 1500 church (the Cathedral of Siena)[21] that from Darmstadt is much far away. […]. These virtual images are not only games for kids but people realize that they are actually learning something and, most of all, they are doing it with no risk at all: just think of pilots who use virtual reality to learn their job, or to physicians who have virtual patients to cure. […] Through the same method, it is also possible to see pathologies virtually (think of TAC examinations, etc.)[22]. These tools have great importance also for history of art teachers because they can use virtual reality to ricreate buildings that do not exist anymore but that are described somewhere. With a simple click, everything disappears, if one wants it.[23]

21 The case of Siena's Cathedral is a part of a more wide project: 'With the advent of technological developments in Virtual Reality (VR) and the abundant content and need for the preservation and conservation of cultural heritage, there has been an explosion of Digital Heritage (DH) projects worldwide. Virtual Reality technology provides an important educational tool to recreate historical content in an immersive high-quality 3D environment where the user can enter and explore the virtual representation in real-time'. The catalogue *Cultural Heritage* provides the 'Peranakans' example: 'Using this VR technology as a base, we have chosen the "Peranakans" specific to the Singapore region as the cultural heritage topic'.

22 Among the examples of 'mixed reality' showed, there is the arthroscopy one, where all operations made by the user to the knee of a dummy are showed on a screen: as if it were a mouse, the instrument the visitor holds in his hand, reproduces the movements in the virtual knee represented on the screen..

23 The interview is taken from a cd presenting the activities of the Cybernarium, which has been kindly sent to me by the internal research staff and translated from German by Ms Raffaella Bo.

In other words: the idea that the funny stereoscopic glasses with polarizing lenses giving a three-dimensional effect could 'become a second pair of sunglasses and could be the same as having a mobile phone'.

3.2 The illusion of immersion: virtual reality of the PercRo

Let's start from the acronym: PercRo stands for 'Perceptual Robotics' and it's a laboratory of the Scuola Superiore S. Anna in Pisa. The field of application of virtuality to art it's not a prerogative of the Cybernarium exclusively: in 2002 the laboratory has presented the project 'Museum of the pure forms' in the hall of Cultural Heritage in Venice, developed with other European partners[24], with the aim of exploring and offering new paradigms of interaction for the fruition of sculpture works. Whereas in traditional museums the visitor experiences the condition of observing the works with no possibilities to interact with them, thanks to the development of aptic interfaces[25], we recover a primary task of sculpture: thanks to touching it is possible to catch the concept of spatiality that the artist wanted to fix in the shapes he has built. The only possibility to see a sculpture is a strong limit[26] (even if we are by now used to this...), compromising a partial fruition for whom has not sight problems, but a total fruition for the blind.

The head of the 'Pure forms' project is Mr Franco Tecchia, an engineer I had the chance to interview some time ago, in order to ask him to explain their *work in progress* after the project launch:

> Art has an important value because poses huge practical problems, such as, for instance, the need of not loosing details in the process of transformation from reality to virtuality. Furthermore, modern museology must face logistical

24 University College London, 3D Scanners UK Ltd, Uppsala University, Centro Gallego de Arte Contemporaneo, Opera della Primaziale Pisana and Pontedera & Tecnologia.

25 The word 'aptic' comes from the ancient Greek 'hapthestai', that in the ancient Greek world used to define the perception of objects and space, through the use of the body as a tactile receptor of information.

26 Personally, in the list of the objects I would like to touch most, there are some works by Bernini and Canova, on exhibition in the Villa Borghese museum in Rome.

> problems: just think about the exhibition on Pharaohs, presently open in Venice. It is not always possible to have all works one would wish to. A reliable digital representation can solve part of these problems.
> The digital has the advantages we already know: first of all, it has an easy maintenance and filing. Finally, one of its main fields of application regards restoration: to plan, for instance, the restoration of a statue through the simulation of hypothesis; to attempt the application of virtual interventions in order to get the expected results. These are all important elements which can constitute a real help.

From a more technical point of view, the results obtained encourage us and come from

> the scanning and the processing of digital information – according to a *bump mapping* algorithm, adjusted in our laboratories – where our first models have been created. On those prototypes, we are bringing the two main virtual interaction modes to convergence: the tactile one and the visual one. At present, the total scanning has been performed on some 'local' works as for Nicola Pisano's 'Madonna col Bambino' and for this project other scanning processes are forthcoming, especially for works coming from museums involved in the project. We are then about to finish a *plug-in* for the web. The results seem to be quite positive, because the graphic effect is very good.

As for the possible fields of application, according to Tecchia,

> the stereoscopic visualisation finds many applications in the scientific and in the industrial branches, especially in the simulation of events and processes. Aptic interfaces, as the exoskeleton, can interest rehabilitation medicine – both motor and cognitive – whereas the *bump mapping* algorithm, laying on the base of the processing of virtual information, has applications in the geographic visualisation, particularly in reference to the GIS. I wish to specify, as a conclusion, that the aspects I am talking about, already used to have a certain independent development and some specific features: the originality of the project 'pure forms' lays rather in the attempt of mutual interaction and development of the individual research fields.

3.3 Immersion without illusion: the 'dialogue in the dark'

We have talked about the importance of the tactile experience and, talking about 'first hand experiences', I would like to tell about one which happened to me, at the end of November 2003. I was in Munich

(Bavaria) for the ECSITE (European Collaborative for Science, Industry and Technology Exhibitions) annual conference and Matteo Merzagora suggests me to try the 'Dialogue in the dark', which was the last piece of a very articulated exhibition brought all over Europe. It's a question of entering in the darkness of an environment which perfectly simulates a real situation. It is necessary to switch off mobile phones there, in order to avoid distractions and lightings, that the eye of a blind person could never experience. The environment recreated in the Munich exhibition was that of a bar: the experience started entering the bar with the aid of a white stick for blinds – which is useful in order to avoid hurting your face against hard obstacles – a very kind blind person would welcome you and ask you if you would like something to drink, if you would like to have a friendly chat or sit down, etc.

I was welcomed by Monica (or maybe Monika), a girl with a sweet voice with whom I communicated in a mixed language of English and Italian. We introduced ourselves and she told me I could trust her within that environment. She initially lead me by the hand but just after that I learned almost immediately where her voice came from. In the bar there was, in fact, a bit of a lively noise and bustle (obviously, as in a real bar, the customers were many and not only us – blind people and non-blind – who were chatting) and music on the background. We walked towards the barman for a drink: I asked him a glass of wine and the barman, blind, of course, kindly asked me if I wanted my wine red or white. I had not specified this detail on purpose, because of the stupid prejudice of presuming that a blind person would certainly not distinguish colours. I chose red wine and Monika did not drink anything because she said that if she would accept all drinks offers from all the exhibition visitors, she would be constantly drunk. I heard the sound of the glass on the bar and, keeping on talking, I tried to understand where it was placed exactly by touching the bar surface with my hand. I was able to find it almost immediately and to understand that it was a chalice: I was able as well to grab it and felt the first real perceptional confusion: it seemed to me as if the distance between the glass and my lips were very long, but it was just a matter of finding the right direction. The time for paying the bill came. The bill was three Euros fifty. I barely knew where my wallet was and I didn't even think about taking my coins out: who can recognise coins in the dark? The problem was that I didn't know how many and what kind of Euro notes I had: I just took one at random and the barman told me 'Ah, a fifty Euro note!', and I didn't even know how to justify my 'choice'. He was probably used to that and just put down my change on the counter, trying to make some little noise for me to understand where it was. I picked up all the money and put it into my wallet and then Monika and me went again to sit for some more conversation: Monika originally comes from Bulgaria but has been living in Germany for quite a long time. I asked her the reason why she knew a little bit of Italian and she told me that she had spent some time near Siena, so we started talking about Tuscany, Italy and other things. I was getting used to the darkness and I wasn't disturbed by it any longer: my sight was 'swiched off' making myself use all other senses, which seemed to increase. It was getting easier for me to move and do things in the dark, everything was getting almost effortless but, unfortunately, the time to go was quickly approaching. I said goodbye to Monika and after she had accompanied me as far as 'the door of darkness', I found my way out.

This experience had left me dumb: this is the 'ground zero' of the immersion, I thought, in the opposite hemisphere of the senses hyper stimulation we are normally subjected to.

4 Conclusions

To tell my story of 'Dialogue in the dark'[27] is useful to think about how it could seem paradoxical the riverse proportionality between sensorial stimulation in an artificial environment – rich in special ef-

27 This experience comes from Andreas Heinecke, that in 1983 works as a journalist to Sudwestfunk's radio in Baden-Baden: here, for the first time, he knows blind people because his task in radio is co-operate with young journalist with this deficit. Two years late, he leaves journalism and works out a project that exploring possibilities of integration for blind people. From these attempt to integration, having clearly in mind the path of sensibilization, born 'Dialogue in the Dark'.

fects – and this, brought to reality, where an important sense as sight is missing.

From an 'added' reality, it is then possible to come to a 'subtraction' reality: if it was possible to write a story of the exhibits, of the experiences made in science centers and museums, then it would be necessary to declare the aims that each of these experiences should have: even though the aims of the 'Dialogue in the dark' are very different from the ones described for the virtual reality, in that case it is possible to find (again) this direct experience, beautiful, enriching us as individuals and maybe, in the specific case, it could also contribute in making us better individuals.

Acknowledgements

I wish to thank Francesco De Ceglia, Matteo Merzagora and Paola Rodari for the contribution to the historical part concerning museums and a lot of bibliography on exhibits; Irene Ferrari for some important linguistic specifications, Elisa Querci and Valeria Romano for the translation of the text into English.

Bibliography

L. Celi, ' 'Anatemnein': note per una storia iconica e semiotica del corpo umano', in *Le Apuane*, XXV, n° 47, 2004, pp. 99-111.

K. Čapek, *R.U.R. Rossum's Universal Robots*, Synergon, Bologna 1995.

H. Coxall, 'How language means: an alternative view of museum text', in G. Kavanagh (ed.), *Museum languages: Objects and texts*, Leicester University Press, London 1991, pp. 85-99.

J. Durant, 'Science museum, or just museum of science?', in S. Pearce (ed.), *Exploring Science in Museums*, Athlone, London 1996, pp. 148-161.

R. Gregory, 'Public perception of science', (editorial to) *Perception*, vol. 29, Pion publisher, London 2000, pp. 1273-1278.

H. Hein, *The Exploratorium: The Museum as Laboratory*, Smithsonian Institution Press, Washington DC 1990.

M. Jeannerod, 'A Neurophysiological Model for the Directional Coding of Reaching Movements', in J. Paillard (ed.), *Brain and Space*, Oxford University Press, Oxford 1991.

G. Kanizsa, *Grammatica del vedere. Saggi su percezione e gestalt*, il Mulino, Bologna 1980.

V. La Paglia, *Il '600: magnificenza del giardino francese e sua influenza nel resto d'Europa (prima parte)*, *Prometheus*, II, n. 32, 1991.

J.-B. Lamarck, 'Mémoire sur les cabinets d'histoire naturelle, et particulièrement sur celui du jardin des plantes' (1790), in M. Landrieu, *Lamarck, le fondateur du transformisme*, 1909, pp. 42-51.

B. Lord, D. G. Lord, *The manual of museum exhibitions*, AltaMira Press, CA 2002.

M. G. Losano, *Storie di automi. Dalla Grecia classica alla Bella Epoque*, Einaudi, Torino 1990.

S. Macdonald, 'Exhibitions of power and power of exhibition', in S. Macdonald (ed.), *Politics of Display. Museums, Science, Culture*, Routledge, London 1998.

M. P. McManus, 'Making sense of exhibits', in G. Kavanagh (ed.), *Museum languages: Objects and texts*, Leicester University Press, London 1991, pp. 35-46.

F. Oppenheimer, 'Everyone is you… or me', in M. Quinn (ed.) *Sharing science: Issues in the Development of Interactive Science and Technology Centres*; Nuffield Foundation in Association with the Committee on the Public Understanding of Science, London 1973.

F. Oppenheimer, *Exhibit conception and design*, talk presented at annual meeting of International Commission on Science Museum, Monterey, Mexico 1980.

E. Pasquinelli, *Introduzione alla percezione aptica*, conference at University of Pisa, Philosophy Dept., 26th June 2002.

E. Pasquinelli, 'Oggetti e presenza in realtà virtuale', in *Sistemi Intelligenti*, XV, n. 3, 2003, pp. 475-491.

M. Van Praet, 'La science en scène', proceedings conference *Information et communication scientifique et technique*, Palais de la Découverte in 1992-1993, Presses de l'Ecole normale supérieure et Palais de la découverte publisher, Paris 1996.

J. Von Schlosser, *Raccolte d'arte e di meraviglie del tardo Rinascimento*, Sansoni, Firenze 2000.

J. Wagensberg, 'In Favor of Scientific Knowledge: The new museums', in S. Lindqvist (ed.), *Museums of Modern Science*, Science History Publications, USA 2000, pp. 129-138.

J. Wagensberg, *Basic principles of modern scientific museology*, talk presented at annual conference of ICOM (International Council of Museums), 2001.

J. Wagensberg, *Emotion for immersion*, talk to the annual conference at ECSITE (European Collaborative for Science, Industry and Technology Exhibitions), Munich 27-29th November 2003.

M. Xanthoudaki (ed.), *A place to discover: teaching science and technology with museums*, Fondazione Nazionale della Scienza e della Tecnologia 'Leonardo Da Vinci', Milano 2002.

Luciano Celi, International School for Advanced Studies (ISAS), Trieste

Gianugo M. Cossi

The user of contemporary art

Abstract:

The study focused on some pavilions displaying technological works of art at the Venice Biennale in 2003, and on a couple of other technological art exhibits. The purpose was to analyze, in the present art-users of this specific field, the motives that guide their choices during the visit, and – at the same time – to understand which modes of fruition characterize this category of users, differentiating it from other types of public. The design of questionnaires answered the need to study the behaviour of the contemporary art users who are interested in the field of technological arts ('artificial art'; videoart; installations; interactive art; videoprojections), in order to build an updated theoretical model of the art user.

The thesis is that the public of technological art is composed of varied types of users, as regards both cultural interests and professional knowledge. Considering that the system of contemporary art, both in institutions and in aesthetic contents, has been strongly influenced by the trends of fashion and advertising, the interpretation of results can be schematized as follows:

1) *the public seems to have adapted to the 'mediatic presentation' of the modern artistic product, which is often displayed in the socio-cultural context of the 'event';*
2) *at present, this trend of the art market seems to be related to a behaviour of the users, which appears to be markedly ritualized (socialized and politically correct, as in all social events), and, at the same time, disillusioned; so that*
3) *the public – in the modes and duration of fruition – can be described at the same time as 'concerned' and 'unconcerned' with the motives guiding the 'consumption' of the art event.*

Therefore, sensible effects of cultural acquisition among users during the visit seem to be rather accidental. On the contrary, such effects were commonly and for a long time believed to take place within the ritual of interaction between visitors and works of art, and in the processes of socialization of socio-visual values traditionally preserved by the artistic heritage.

Keywords:
electronic art, video art, technological art, galleries, museums, methodology, public, education, employment

1 Scientific issues

Electronic arts are those arts involving new technologies; that is to say, produced first with technology from the electronic age, and then evolved thanks to computer civilisation. The expression derives from the Corso di estetica (1989) by Renato Barilli. Those goods available on this market are spread around an art gallery circuit that is well-defined, autonomous and characterised by international acceptance. Italy, although largely without specialised art galleries, has attempted to compensate for its marginal condition by adopting courageous initiatives especially at famous galleries. The most advanced countries in this sector are: the United States, Germany, the United Kingdom and Japan. Research has been mostly involved with the electronic sector, being the most analysed and widespread, as well as video icon art (or info icon art) including video art, video installations and video projections.

2 Principal research orientation

These achievements have been analysed by literature in the sector as a sort of coordination between heterogeneous technological practices, used in the fields of production, fruition and aesthetic interpretation. According to the fields of study, the different genres have been defined as some sort of 'hybrid' projects that have matured in the orbit of technical sciences [Popper, 1993], or considered to be 'machines of imagination' [Luhmann, 2000], that is to say, inherent in the field of social communication more than that of art values. For L. Qvortrup instead, it becomes appropriate to speak of an aesthetic re-definition (that shifts the value of the good; from that of its commercial use to a gallery use) of the multimedia means [Qvortrup, 1998]. In order of their historical appearance, technological art tendencies studied by the principal researchers are classified into: video art (video projections and installations), computer art, computer graphics, visual effects, net art, cd-rom art and virtual art.

3 Methodology of the study

This study was conducted between the summer and winter of 2003. Several hundred different users were interviewed in various Italian cities, in an early stage in Monfalcone (in the province of Gorizia) and at the Venice Biennale in the middle of July. The second tranche began in Trieste at the end of October, and concluded in December 2003 by returning to Monfalcone[1]. Use was made of interviews through a partly structured questionnaire, administered individually (with a face-to-face technique) to a casual sampling of users at the end of their visit. In comparison with in-depth investigations, the more general nature of the contents in a semi-structured questionnaire is partially compensated for by the possibility of the pollster to insert a flexible number of questions and reach a conveniently vast sampling in a rather short period of time.

This study of the public must be collocated in a general analytical framework of museum demand. Generally, a visit to the museum is characterised by its exceptionality. In the major art museums, galleries included, the logic of occasional convenience involving small groups (of three to twenty participants) seems to prevail. Those who enter museums during a trip profit from a previous visit by a friend or relative. The museum public is therefore opportunist and only rarely composed of isolated users.

1 The research was conducted at the *Monfalcone Gallery of Contemporary Art*, and in the Dutch, Canadian and Israeli pavilions during the *Venetian Art Biennale* 2003, as well as in the *Studio Tommaseo Gallery* in Trieste.

4 Characteristics of the public

The initial questions are the classical structural ones[2] [Pitrone, 1984]. Classified into such categories as: habitual user, occasional user, curious visitor[3], the public for electronic art is made up of many young people, between the ages of 26 and 35 (27.3% of the total). That age group loosely called *young adult*, more or less between the ages of 19 and 25, is nine percentage points below the former category (16.3%). The highest number of visitors includes those who are more than 36 years old (51.8%), even though the composition of this age bracket is not homogeneous. For practical reasons which emerged during data processing[4], there is no distinction between, for instance, fortysomethings and elder groups. The public was largely composed of males (60%; Table 1), while at an educational level, the most significant presence was those of secondary school graduates (59.5%, Table 2) followed by university graduates (28.5%) and middle-school graduates (12%). There was a significant presence of pensioners, while the presence of government employees was minimal. From literature on the sociology of the public for art and museum marketing, we can deduce that, in effect, secondary school graduates constitute that category most oriented to visiting specialised museums [Avorio 1999, p. 54] and well-known galleries.

Table 1 – composition of the public by gender

Gender	Frequency	Percentage
Female	160	40.0
Male	240	60.0
Total	400	100.0

2 Called thusly because they regard those fundamental attributes, pillars of the interviewee, such as: place of residence, age, gender, degree of education, etc.

3 The three categories which were used to classify the users were constructed on the basis of the declared number of visits in a calendar year. The user, before responding, was informed that *curious visitor* indicated not more than 1 or 2 visits per year to an electronic art exhibit, *occasional user* indicated from 3 to 6 visits and *habitual user* 7 or more visits annually.

4 It has been noted that, as age increased, this variation bore no influence at all on the degree of significance of the answer.

Table 2 – educational level of the public

Educational level	Frequency	Percentage
Compulsory schooling	48	12.0
Secondary school	238	59.5
University	114	28.5
Total	400	100.0

Interesting data is obtained from the context of the job/professional occupation of the users. Full-time employees (Table 3) were the most represented (39.8%). These were followed by self-employed professionals (21.3%), university students (14.5%) and pensioners (6.3%).

Table 3 – socio-professional level of the public

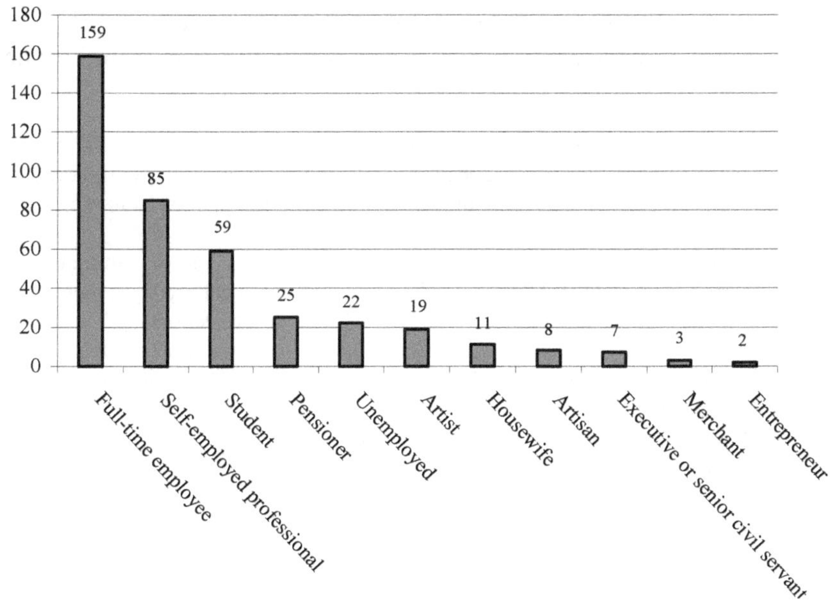

Rather minimal were the presences of artisans (2%), senior civil servants (1.8%) and, a remarkable statistic, artists (4.8%) as a whole, a heterogeneous category including the multi-faceted circle of technological artists.

5 Study of the user responses on the basis of the social theory of the media

We assume the postulate that the public for electronic art is part of a sub-population of consumers of industrial sub-culture, that produced by the mew media. The importance of the methodological habits for media consumption seems to constitute an ineluctable premise for each type of study aimed at examining the functioning of the consumption practices of technological aesthetics[5].

The planning of Global Culture, in part predicted by the theories of E. Morin [1969] and A. Moles [1967][6], responds to optimised criteria of distribution and semantic re-formulation, for goods that owe to managerial and supra-national centres[7], not only their reason for being (as products) but also their reason for reacting (as related media)[8]. From the market for products we would thus achieve a market for trademarks[9], in which the user is often *educated* to re-launch their ac-

5 Among the first, Harold Innis had explained that it is the use of media, and not their simple production (while appearing to be a new technical support structure), that engineers those phenomena of transformation in the field of communications and culture [Innis, 1982].

6 This refers to Edgar Morin's sociological theory of mass communication (*L'industria culturale*, 1969) – according to which mass culture corresponds to a mythological form, made up of 'a system of specific projections and identifications', and to Abraham Moles' cybernetic study of socio-cultural production (*Sociodynamique de la culture*, 1967). For Moles, the knowledge induced by that industry gave to the masses a 'mosaic' of sensations, where it was no longer possible to discover any fundamental idea, but merely ideas – command, assertive or 'key words'. Combinations of words that are highly significant for mass imagination.

7 An objective aiming to favour the increasing dimension and complexity of endeavours '[...] is that of exploiting economies of scale and reciprocal dependencies to offer a complete packet of information technology: computers, telecommunications, electronic goods and components, etc.' [Kumar 2000, p. 43; translation provided by the author].

8 'Advertising puts the relationship between man and the world on a stage, (this relationship) made up of attractions and repulsions [...]' [Montes 1998, p. 43; translation provided by the author].

9 The devisers of the majors move therefore on the different fronts of communication and economical and political partnership, so that their devices develop, in re-

tive primary role through unprecedented discretion – ambassadorial yet creative – assumed together with the product it wears, or of which it speaks. The context is separated from the classical, bipolar conception of the economy, which includes two uniform fronts: producers and consumers. Thanks to marketing, and to the technology for the diffusion of seductive advertising language and market propaganda, as Bruno Sanguanini writes, the system of the new media[10] is not only a complex definition, but also, consciously, offers according to a rather weak means identity,

> from the moment that the multiplicity of the structural components is so strong as to constitute a sort of multiple identity [...]. The role of the 'canal' crucial to the definition of the mass medium is substituted by the system of the various structural devices distinguished more by their technological aspect rather than their cultural genre [Sanguanini 2003, p. 42; translation provided by the author].

Following the possibilities presented by the versatility of the new media, the cultural implications appear significant, both on a level of the expectations toward the world of information (which should have a practical and alluring interface; and offer data easy to download), that (b) regards what we mean when we speak of the nature of 'knowledge'[11].

Let us take into consideration the first point. Surely, an effective metaphor to summarise the expectations and hopes that society nurtures with regard to itself, to its current development and to its future in the direction of a progressive perspective, might be that of science fiction. This has increasingly become a sort of ideology of the new

---

tailing, features defined as personalizing. The sociology of legitimation of tastes has studied thoroughly the many homologating and – at the opposite – fragmenting aspects of the forms of cultural consumption [Bourdieu, 1983].

10 From which the author offers again [from Garassini, 2000] the definition: 'the combination of new means of communication constituted by Internet, digital TV, multimedia support (of any type, form or function)' [Sanguanini 2003, p. 40; translation provided by the author].

11 To the point that between consciousness and comfort, in the objective of many professional advertisers, it would be convenient to juxtapose formal and functional equivalencies.

human consciousness, invented by the industry of serial narration. And yet, for a combination of reasons, the relationship between technological art and science fiction is not as immediate as it might initially seem[12]. In any case, the public has judged the suggested link between electronic art and the commercial sector falling under the label science fiction as largely legitimate and justifiable. Some 272 users (equal to 68%) confirmed the probable existence of an analogical nexus between the technophile utopias of Asimov & Co. and video art. The largest number of supporters of that link are the curious users (124, equal to 45.6% of those who answered 'yes'). One of the possible interpretations of this statistic regards the fact that, for a large part of the vast category of cultural event consumers, placing a monitor under the eyes of aesthetic interest is equivalent to transfiguring video art in an aesthetic symbol[13] of the ease in consulting information.

This could be linked to the demand on an informative level that regards the second point above (nature of knowledge). That is if, by average use, the hypothesis is valid that this represents a target of cultural acquisition, today the exigency appears to depend more and more on the expectations of daily decisions, to be verified in the maximum condition of its ease of use[14], typical of the behaviour of the consumer. As said, the pairing – wealthy society and electronic art – seemed to us to be worth exploring, presenting various risks but, at the same time, recurring links and limpid implications (Table 4). In this sense, the link between the wealthy society and electronic arts has been studied.

12 Not only because much video-art, for example continues to offer anachronistic canons, from that of a militant underground culture (something that even the non-specialist public senses) yet also because the contents themselves of the video-films and projection do not usually regard subjects treated in the better known spheres of fantasy literature.

13 The excess, of the unjustified nature of those processes of symbolism is often the social symptom of a tendency towards obsession and 'sensory distraction' [Buck – Morss, 1992], but also towards a glamorous affectation to pose as an aesthete.

14 Facility in the association between concepts; or between concepts, goods and *social memory*.

Table 4 – intersection between degree of agreement with the affirmation that electronic art 'represents wealth' and the variable 'category of user'

Degree of agreement		Category of user			Total
		Habitual User	Occasional User	Curious	
'Is Technological Art the product of a wealthy society?'	Not at all	15 (12.4%)	8 (8.1%)	20 (11.1%)	43 (10.8%)
	Somewhat	19 (15.7%)	22 (22.2%)	35 (19.4%)	76 (19%)
	Rather true	50 (41.3%)	50 (50.5%)	89 (49.4%)	189 (47.3%)
	Very true	37 (30.6%)	19 (19.2%)	36 (20%)	92 (23%)
Total		121 (30.3%)	99 (24.8%)	180 (45%)	400 (100%)

As we see in Table 4 the majority of the habitual visitors (87; equal to 71.9% of the category) and the occasional visitors (69, equal to 69.7% of that category) and the curious (125, equal to 69.4%) recognised a complete sense of correlation. The public recognises in the hybrid pairing – video plus art – the same typical mix of environmental and material culture in which we collocate those well-known activities related to wealth, intended as easy consumption. Articulated by the use in the reifying organisation of time[15]. The creation of the aesthete through the monitor reminds the 261 favourable users (65.3% of the total) that post-modern society is difficult to separate from the emotional menu offered by technologies for the home and office, nurseries (these latter) of a significant 'value added' constant in life. To conclude on this aspect, we can say that: the average user *observes technological art using judgment and comparison parameters perfected through years of New Media consumption.*

Through the diffusion of the New Media, as Danila Bertasio writes, not only do we act on a complex system, on museum organisations, on the entire art system, but also, more generally, within the present di-

15 M. de Certeau's studies on cultural practices places the accent on the diachronic role which, in daily life, is assumed by various common objects, such as the paperweight or the television. Functional, in and of themselves, they may become time 'activators' for a world of users that is more and more distinguished by a stop and go phenomenon, determined by the use of more banal goods of exchange. In this commercialisation of time also, there exists a plurality of material culture that determine the physiognomy of daily life [De Certeau, 2001].

mension of an ontological focus. Computerisation influences the way that we pursue knowledge given that it is more and more 'the majority of us, sensing the difficulties of intellectual work, tend to prefer a version of "knowledge" that, avoiding the stages of its production, arrives rapidly at its conclusions' [Bertasio 2003, p.44; translation provided by the author].

6 Motivational analysis

To construct a model of the average user it is necessary to refer to motivational research. We proceeded by degrees. Some 273 users like the exhibit, equal to 68.3% of the total. Among those favourable, the very satisfied were more or less thirteen percent of the total (12.8%) while, on the other hand, the discontent were around 30 percent (31.8%). In the successive stop, from the intersection between the variable education and that of agreement, we note that the differences in cultural preparation do not significantly influence aesthetic judgment. The fundamental divisions that influence how judgments are made, we shall see, are of another nature. Instead, what is interesting is that the preferential evaluation is made explicit with a measure of sufficiency. The exhibit was rather appreciated, so an average positive value was granted to it (55.5% of the user total).

To understand the motivational field, we use a functional configuration of 'field of strengths' [Bourdieu, 1972][16]. The users were asked why they came to the exhibit. So as to prevent purely mechanical answers the interrogation was posed as an open question. The attitude of the interviewer was focused on the dialectical methods of the user, to limit the risk of answers that were hardly personal or involving[17].

16 The limits of that field frame the artistic system, made up of symbols, goods, social actors and worlds devoted to allocation. In this study we affirm its autonomy to assist its differentiation from other systems used to communicate culture.

17 We wanted to capture the full attention of the interviewee so that they would enunciate, while filling in the written form at the same time, some sentences (even if banal) that would show an effort at subjective interpretation (for example) of the aesthetic value of the art on display, thus responding to criteria of *directive* methodology. As was demonstrated by studies that used (in this case) the tech-

From the answers obtained (with the system of grouping by recurring, prevalent themes) we notice that the largest part of the public (100 users, equal to 25% of the total) headed for the exhibit out of a desire for pure diversion (diagram 1). The second most numerous category (81, equal to 20.3% of the total) was made up of those persons who have a passion for innovation, either technological or cultural. Globally, some 45.25% of the users (the total percentage of the first two categories) declared they were more interested in using their non-working time in an unconventional manner, rather than dedicating it to the evaluation of art, or more formal declarations.

Diagram 1 – the users' motivations for the visit

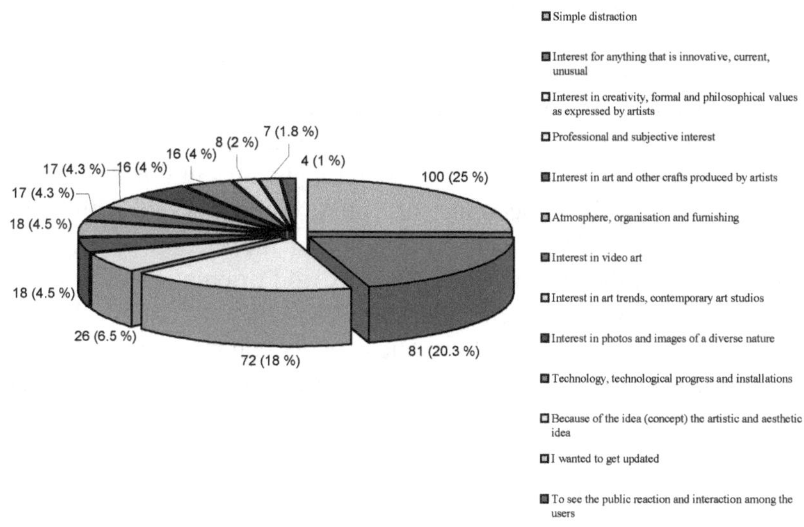

Aesthetic research appears relegated to the background. In fact, just 18% of the users (value that places it at third place in the taxonomic pattern) declared that they motivated their presence as satisfying an

---

nique of verbalisation, developed by Cristina Russo and Bianca Vasta [Russo and Vasta, 1988] the use of non-standard, matched multiple techniques often produces good results for the increase of public awareness.

interest in formal and philosophical values expressed by artists. In short, an interest in distraction or participation according to conditions of diversion seem to constitute the principal motive for the average user that crosses the threshold of a gallery or a technological pavilion. Works of art constitute a *strategic occasion*, which can be used in the social rites of communication, a virtual communicational chessboard; the field of strengths made sociologically dynamic, distinctive by the second order observation. In this study, the observer becomes *user*, in the moment in which they 'nominate' (virtually) someone to produce distinctions.

7 Changes in appreciation according to the cybernetic theory of the Second Order

In order to analyse the comments expressed on the electronic art, we use the social theory of averages. So as to complete the construction of the theoretic profile of the average user, that is, to interpret the relational practices of the public together with changes in one's own criteria for appreciation, it becomes useful to equip oneself with second order cybernetics. The social importance of the second order, in particular as far as the effects of conformity and orientation of collective attitudes is concerned, is considerable. This a social result which can not be treated as a universal law. The process can be applied whenever a group of potential visitors can not (or have no interest in) producing critical evaluations on a work of art, nor on the communicative characteristics of the public with which they come into contact[18]. In this case, the absence of discriminating intention produces impoverishing aesthetic effects, with the result that it is the presence of the public which constitutes one of the principal objects of participatory investigation. In particular, in its internal component, we distinguish those authorised to produce distinctions, those that are distinguished by others to orient the observations of those who are not observed. In prac-

18 Including the critics, the guides at exhibits and the experts that appear before the exhibit, on the small screen. The second order can be activated even before a social event takes place.

tice, there are among the factors that serve to change users' minds. In this case, the electronic genre is presented, in the course of the exhibit, as an instrumental projection of the social configuration that communication progressively assumes: initial coldness, awareness of the role of the participants[19], of the scorers, etc.

> *Whoever observes forms, observes other observers in the rigorous sense that he is not interested in the materiality, the motives, expectations, or utterances of these observers, but strictly and exclusively in their use of distinctions [Luhmann 2000, p. 67].*

In the context of the second order, we can comprehend why, as a result of the comparison between diagram 1 and 2, the public has changed its mind about several fundamental aspects, in particular the importance that it recognises in technological art as a creative fact. In the pattern of the following page we can note a change in the orientation of the public between the 'before' and 'after' exhibit. A fundamental statistic, the importance of the recognition of art as a creative fact has increased. From the 25% of the total users that had expressed their motivation as simple entertainment (diagram 1) we pass, within the category of 273 persons who appreciate the art (diagram 2) to 2% of the 400 users (which rises to 3% relative to the category under consideration).

Given that we are analysing the category 'I liked the exhibit' this result might not be surprising. The fact is that changes of attitude of equal importance were registered in other categories also. Those who were interested in 'creativity' increased also as an absolute value, passing from 72 (18% of the total) to 92 persons (23% of the total; 33% relative percentage). The two categories that had expressed interest in the 'presentation and atmosphere' and any possible 'interaction with other users' (important fact) experienced a considerable increase

19 In the social cybernetic interpretation, art is valid as a conventional plane to facilitate the progression of the encounter. The discursive activity under way (viewing the exhibit; which should constitute, formally, the principal activity) is therefore pre-textual, valid for control and comparison functions that are socially achievable: glances of comparison and intimidation; verifying activities, protection of one's own prestige, etc.

in their absolute values. These grew respectively from the preceding 18 (4.5%) and 4 (1%) to 47 (11.7% of the total; 17% relative) and 46 (11.5% of the total; 17% relative). Above all, the number of references to video art, hardly considered in the initial approach to the exhibit, increased. At the end of the exhibit video art is judged more interesting, passing from 17 users (equal to 4.3%) to 32 users of the group of those who appreciated the exhibit (85 of the total, 12% relative).

Diagram 2 – reasons you like the exhibit
Opinions expressed by those belonging to the category of the 273 users who appreciated the exhibit

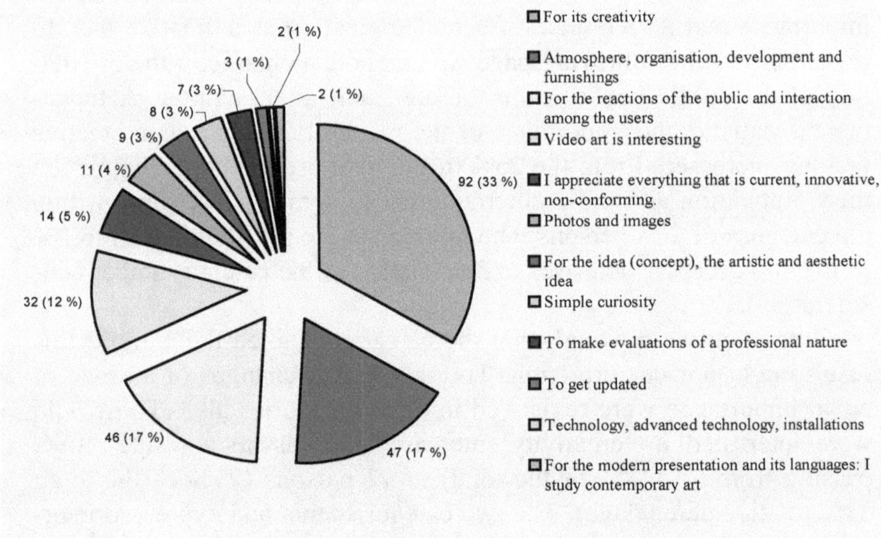

In summation, when studying second order dynamics it is possible to hypothesize how, more and more often, the public is not the human content of the cultural event, but paradoxically, serves instead (according to a social point of view) as a 'functional container' of the same. This can be said for art exhibits lacking an adequate formal specificity, almost uniform in their fragmentary nature. In these cases the public is lead, so as to interact significantly, to create self-references through auto-observation. We could then say: it is the public that identifies exhibits that lack distinctive titles. Which, in a cybernetic sense, transforms them into a sort of container.

8 Conclusions

We recognise that the experience of the 'visitation' oriented the majority of the users in a strong sense, towards a positive observation, participating in the aesthetics. Art, or the interest in creativity, which, initially did not seem very considered, earned dignity, as well as, during the socialisation stage, criteria of distinction and judgment, such as to report – as would seem correct – art (or what is said on this) in an acceptable condition of appreciation. Is socialisation what redeems aesthetics? Is it not strange that the public recognised in interaction factors, and therefore in discursive activity engaged in with other users, one of the reasons for interest in the exhibit in which they participated.

It has been said that the average user observes technological art using evaluation and comparison parameters that have been honed in years of multi-media consumption. The results of the study, while proposing a sort of identity between the methodology of new media fruition, and the methodology of the users' fruition of the exhibits considered, sought to discern one of the principal reasons for the success of each form of communication that intends on promoting the artistic phenomenon socially; the importance of social rites, the importance of multimedia literacy. Which is synonymous with recognising, as written by Raimondo Strassoldo that the mechanisms that involve the artistic system are

> the same fundamental mechanisms that characterise each socio-cultural evolutionary process. The two fundamental mechanisms are that of functional differentiation specialisation, which generates differences; and that of the recombination of these differences into homogeneous units, but at a higher level of complexity [Strassoldo 1998, p. 67; translation provided by the author].

In the past, the difficulty of aesthetic appreciation, implicit in the organic and hierarchical classification of culture, constituted one of the factors that discouraged inter-class participation in art exhibits. At present, no one performs the role of dissuader, given that the new culture does not (necessarily) require a severe apprenticeship, a change of route, a re-thinking matured in the hope of new acquisition of knowledge. Currently, it is certainly not necessary to possess a special artistic theory to feel rightfully included in the public of contemporary art. On the contrary, interest in the system of art promotion seems to act with far-sighted wisdom. The objective is that the average user responds to a type of all-inclusive cultural initiative, aimed at involving the adequate communicative contexts, especially so as to promote the exposition of art as a test of each persons' ability in social diplomacy. The interest of the promotional system is that the average user should respond to a type of cultural supply that is theoretically rather ample, aimed at involving adequate communicational spheres in the realm of cultural tourism.

Bibliograpyhy

A. Avorio, *Il marketing dei musei*, Edizioni Seam, Milano 1999.

R. Barilli, *Corso di estetica*, Il Mulino, Bologna 1989.

D. Bertasio, *Solitudine e condivisione dell'arte*, Liguori Editore, Napoli 2003.

P. Bourdieu, *L'amore dell'arte, le leggi della diffusione culturale: i musei d'arte europei e il loro pubblico*, Guaraldi, Rimini 1972.

P. Bourdieu, *La distinzione. Critica sociale del gusto*, Il Mulino, Bologna 1983.

S. Brancato, *Sociologie dell'immaginario. Forme del fantastico e industria culturale*, Carocci, Roma 2000.

S. Buck-Morss, 'Aesthetics and anaesthetics: Walter Benjamin's artwork essay reconsidered', in *October*, vol. 62-*fall 1992*, October Magazine Eds., NY 1992.

M. De Certeau, *L'invenzione del quotidiano*, Ed. Lavoro, Roma 2001.

E. J. Hobsbawm, *Il secolo breve. 1914-1991: l'era dei grandi cataclismi*, Rizzoli, Milano 1995.

H. A. Innis, *La tendenziosità della comunicazione*, Sugarco, Milano 1982.

K. Kumar, *Le nuove teorie del mondo contemporaneo. Dalla società post-industriale alla società post-moderna*, Einaudi, Milano 2000.

P. Lévy, *Cybercultura. Gli usi sociali delle nuove tecnologie*, Feltrinelli, Milano 1999.

N. Luhmann, *Art as a Social System*, Stanford Univ. Press, Stanford CA 2000.

A. Moles, *Sociodynamique de la culture*, La Haye, Paris 1967.
S. Montes, 'Stili di vita nelle pubblicità di due marchi di sigarette', in G. P. Caprettini and G. Ferraro, *Leggere la comunicazione. Politica, Pubblicità, Internet*, Meltemi, Roma 1998.
E. Morin, *L'industria culturale*, Il Mulino, Bologna 1969.
M. Negrotti, *Artificiale. La riproduzione della natura e le sue leggi*, Laterza, Bari 2000.
F. Popper, *Art of the Electronic Age*, Thames and Hudson Ltd., London 1993.
M. C. Pitrone, *Il sondaggio*, Franco Angeli, Milano 1984.
L. Qvortrup, 'L'arte e la società dei multimedia interattivi: lo shaping socio-estetico dei multimedia', in Bertasio D. (a cura di), *Immagini sociali dell'arte*, Dedalo, Bari 1998.
B. Russo and C. Vasta, 'Uso combinato di scale Likert e figure tematiche in una ricerca sugli atteggiamenti', in Marradi (a cura di), *Costruire il dato*, Franco Angeli, Milano 1988.
B. Sanguanini, 'New Media e teenager: Tamagotschi, Game Boy, computer e videogames in Italia', in L. Bovone and L. Ruggerone, (a cura di), *Intorno a produzione e consumo. Percorsi nell'industria culturale*, Franco Angeli, Milano 2003.
W. Santagata, *Simbolo e merce. I mercati dei giovani artisti e le istituzioni dell'arte contemporanea*, Il Mulino, Bologna 1998.
A. Semprini, *Marche e mondi possibili. Un approccio semiotico al marketing della marca*, Franco Angeli, Milano 1993.
R. Strassoldo, *Forma e funzione. Introduzione alla sociologia dell'arte*, Forum, Udine 1998.
J. B. Thompson, *Mezzi di comunicazione e modernità. Una teoria sociale dei media*, Il Mulino, Bologna 1998.

Gianugo M. Cossi, Dipartimento di Economia Società e Territorio, Università di Udine

Franco Fileni

Licence of use: the new owners of knowledge

Abstract:

By analysing briefly the way the user is considered, particularly by the means of research which investigate his relationship with the computer and with various other related problems (i.e. the interface), it is possible to reach the conclusion that this concept is not only unlikely to disappear, but will be more and more amplified.

This fact depends upon the realisation that in a society in which the services, and in general the majority of individuals, will make an increasingly widespread and sophisticated use of information, the role of 'user' is destined to develop.

From this point of view, the 'user' is created not just by the difficulties of interaction (interface), but also by the lack of some property of a component of the system possessed by the 'user'.

Keywords:
User, use licence, World Wide Web, interface, usability, bios, communication, brainframe, navigability, interaction.

In the etymological dictionary of the Italian Language of Cortellazzo-Zolli (Bologna 1988) you can read this interesting quotation dealt with by A. De Nino *Frequent Errors still in use in the Italian Language* (Firenze 1866): 'User: is a piece of vocabulary that might be put in the dictionaries of the highest esteem (when they are compiled by those who have the letters where they have the horses)'. L. Del Tutto, in an internal document directed to a group of researchers, starts with a discussion of the term 'user'. In the final part he also presents an interpretation of the obscure horse figure tort in dance and, in the two possibilities that he proposes, a negative attitude is inferred of De Nino as regards this term: 'in both cases the judgement is categorical: user will not have a place in the esteemed dictionaries!'[1]

After 136 year, however, the term appears both in the dictionary of Zingarelli (2003), with the following definition: 'User, 1) Who uses goods or services, specially public; 2) by extension, Who uses a common cultural, social or spiritual heritage' and as an example 'the users of the Italian language' is given; and in the everyday linguistic use where it is used more and more. It is employed not only in the Informatics field but also in the contexts distant to this area (user of the telephone, of the motorway, of public means, of health, of museums, etc.) and to such an extent that its meaning becomes unclear and not well defined. In the final analysis the term 'user' seems to be more used than the alternatives: citizen, customer, person, patient, etc.

All linguistic terms exist

> really only when there are the conditions for their existence, and these are determined by their cultural conditions. That means that one can always make the analysis of a word in abstract, but at the same time there is the need to identify the historical and (therefore) cultural contexts that have afforded its existence and wealth because these are the things that give it real content and a value in the system[2].

The context which is usually referred to for the birth of the term 'user' seems to be that of the advent of advanced technology, especially Information Technology, with the research done by the *Human computer Interaction* (HCI) with specifications slowly but increasingly more contextualised, as will be seen, accompanied by its pervasive advance from Informatics into various everyday contexts. Therefore, from the extended definition of Zingarelli to this last observation, one can deduce that the inflation of the term comes from the fact that the Informatics context has become a kind of common cultural and social heritage and by the fact that it works in other socio-cultural contexts. Somehow the informatics functionalities have informed – in the etymological sense of the term 'to give shape' – also other related envi-

1 L. Del Tutto, ' "Utente", "utenza" e forme europee comparabili', in M. Negrotti (ed.), *Homo utens*, Guerini e Associati, Milano 2004.

2 *Ibidem*, p. 4.

ronments equating the services to the informatics services procedure. Already this should not astonish in as much as the use of communal definitions like 'the information society', 'the informed society', 'the society of the net', etc. The personal computer (PC), born as an instrument of work, takes a central role in the major part of the activities conducted by the individual: with the PC you work, you play, you study, you communicate, you buy, and it puts us in touch with the world. Therefore it seems apparent that the development of the concept of the user is closely linked to the development of information technology with particular reference to the PC. The expression 'end user' is often used, this can be a 'skilled' or an 'unskilled' person, and denotes an individual that posses a good competency in the operating of software (sw), but it is also used, albeit in a very sophisticated way, in the phrase 'software user'. In various degrees the user is located in a *continuum* that goes from artificial to human and that, in various contexts, is associated with and confused with the citizen, customer, etc. and thus becomes more complex. The moment one enters a shop and buys a software product one is adorned with various roles simultaneously: customer in the shop but also user in as much as is asked, for example, from the technical explanation. Also the role of user starts in relation to the company that has produced the sw product (usually in relation to some operation difficulties) and, often, as user of the program, one participates in a forum or in a virtual community thus having a citizenship, in that community, primarily the user of a product.

The concept of user, therefore, reveals two very important aspects: the first is based on dynamism and the second on the relative aspect linked to the context, to the situation and to the point of view of one of the two elements that enter in a dialogue.

In this paper I will try to examine the type of dialogue, in reference to the concept of the 'user' also with some negative connotations that often accompany it, referring explicitly to the context of social informatics and to the world of the PC in a society of networks[3].

3 Cf. M. Castells, *Galassia Internet*, Feltrinelli, Milano 2002 and *La nascita della società in rete*, Università Bocconi Editore, Milano 2002.

Analysing various texts and material and those found on the net, it emerges that the idea of the user, with all the problems connected to the learning of the use of instruments and of the interaction with sw and hw (hardware) – problems which are still present –, started with the diffusion of the PC in the 1980's when the informatic instrument turned from being an 'object of work' to being an 'instrument for work'. Upstream from the birth of the user there is however an expert that does not only know and write sw, but also knows how to examine hw, often built according to his specifications. The instrument seems clear to him, there are no problems of interface in as much as he who uses it is also the constructor. With the PC and with the user, that becomes, or better to say should become, a central figure, start the research to improve and ease the relationship of interaction with the machine.

Downstream we find, however, the user at the undertaking with the advent of the first multimedia and, successively, with the net and the benefit from the WWW. In the future, at present proposed as an experiment, it will compare the… 'disappearance of the user' due to both the new kinds of interface and to an increasingly everyday and personal use of informatic equipment (wearable computers, biochips, computerised household appliances, etc). The idea of the disappearance of the user in respect to increasingly more portable and personal technologies is that they are founded 'totally with the unity of the accessories that we are used to carrying (watches, glasses, etc.) and with all the environment that surrounds us'[4], in reality it seems to me very distant at the moment and, from the point of view of the interaction with the informatic object, maybe completely improbable even if the learning of use and interface problems[5] are completely eliminated, which is a very difficult thing.

4 R. Scalisi, *Users*, Guerini e Associati, Milano 2001, p. 145.

5 The disappearance of a certain part of the user can happen, according to my vision, when the wearable or directly inserted into the body computers allow their use with the utmost of expressiveness not particularly formalised and the minimum of learning use. The remaining part of user will stay until there will the necessity to buy whatever type of software, as I will show in the following work.

The PC encompasses various internal functionalities, previously delegated to other instruments, which have transformed it into a 'personal computer'. This passage has occurred thanks to the development of the networks and, as regards most people, the development of the WWW, the service of the Internet that today takes on to its internal service things previously directed by other programs (via the web post is sent and received, files are taken, films are seen, one communicates with others with audio and video, economic transactions are executed).

The importance assumed by the WWW has in the last years brought another extremely important concept into the limelight connected with the user: usability. In the 60's the problem of usability came into existence within the environment of the ergonomics in the man-technology interaction sector, but today it finds its more specific position thanks directly to the development of the World Wide Web. The ISO 9241-11[6] defines usability as 'the ability of a product to facilitate a specific use in the achieving of specific objectives with effectiveness, efficiency and satisfaction in a specific context of use.' In the web usability is adapted, from ISO, – as in the classic definition that Visciola gives of it 'a Web site is usable when it satisfies the informatic needs of the end user that is visiting it and questioning it, providing ease of access and navigability and reaching a suitable level of comprehension of the contents. In the case that not all the information is available, a good site requests from other informative sources[7].' – for this particular context. But an extremely important point of view changes: in the usability of a generic sw (word processor, design, for calculus, etc.) the user finds himself in front of a product with its interface *after* having bought it, with all that which ensues, while in a web site the user confronts a problem of usability *before* the purchasing of the product[8]. There is a principle relationship client → user, while in a web site we have the relationship user → client much more relevant in a certain way for those that sell.

6 *International Organisation for Standardization.*

7 M Visciola, *Usabilità dei siti web*, Apogeo, Milano 2000.

8 Cf. J. Nielsen, *Web usabilità*, Apogeo, Milano 2000 and *Homepage usabilità*, Apogeo, Milano 2003.

If, in general, the usability of a sw refers primarily to the design of the graphic interfaces (GUI, Grafic User Interface), in the web it assumes an even more important role due to the concentration of services that are benefited via the browser that is becoming 'the sole interface' for the net for most users. User, interface and usability are thus found in an extremely close, important and inseparable relationship.

But how has this particular type of use come about? Does the simple term user define and mark a sole type of relationship or does it encompass many?

Every individual who sits, for whatever reason, in front of a PC engages in various types of relationships that overlap each other or cohabit with this instrument simultaneously, and these are generally determined by the development of the interface, that in the web – as already seen – also poses the concept of 'usability', there is therefore a boundary, the boundary that marks a cognitional-communicative space of the user and another determined by the logic of the hw and of the sw. Such boundaries do not represent an impenetrable barrier[9], otherwise there would not be means of transmission, rather there is the place-time in which the transformation happens from a cognitive system to another: it is the moment of the transformation from one code to another. The will of the user to execute an operation initially translated, for example, in the instruction to move the mouse in a certain position; this analogical process of the body that moves the mouse with the eventual clicking of the mouse becomes, then, harmonised with digital elements that allow the sw to act on the hw in a certain way (highlight, for example, a word in a word processor) coherent to completed movement. The boundary represents the translator from the will of the user but also his constraint: it is possible to execute operations and processes that are permitted by the logic of the program and the structure of the hw. The interface determines, therefore, the relationship of a user with the informatic unity that he is using and specifies his relationship, in as much as the user for example 'skilled' or 'novice', in general, is usually defined as 'end user'. The reposition-

9 Cf. A. Wilden, voce *Comunicazione*, Enciclopedia Einaudi, 3rd vol. and F. Fileni, *Analogico e digitale*, Goliardiche Edizioni, Trieste 2003.

ing, not reduction *toutcourt*, of the complexity of the external and internal (from the interface to the lines of programming that define the possibilities of the program) is the price one pays for a relative simplicity of interaction with the modern PC. Every PC, but also other instruments of an informatic nature, is usually defined as a single unit of hw and sw. It is always thought of last in terms of programs that are used for some precise goal, to elaborate an image, to write, to calculate etc. it is in respect to this part that the individual becomes a user: the end user, with all the degrees from expert to novice, of a sw which simply holds the licence of use. It is only in the comparison of hw that user disappears and the 'owner' reappears: the manuals of use of the programs are titled as 'User manuals'; the manual that arrived at the same time as the computer with which I am writing these pages is titled however as 'Owner manual'. I own a physical object but I only use a sw; the one without the other is useless, it seems however that the symbiosis of the PC breaks also the individual into two entities, owner and user, in an impartial way, but there is still a level to explicitly express. Together with the machine that is bought we always find a particular sw installed defined as Bios[10] thus that sw that prepares the physical aspect of the machine at the installation of the operating system, for example Windows or Linux, that then lets us use other programs. As with all sw, also the Bios does not belong to the person that bought the PC but they only have a licence of use; therefore if I do not become a user I am an owner, but simply of a reasonably good-looking and cumbersome knick-knack. The idea of 'user' thus seems to be dominant in front of whatever informatic instrument over the idea of 'owner' and also that of client. Buying a sw or a PC via a retailer the individual receives a materiality, the box with the cd-rom or hw, manuals etc., and it is in that moment the client, both of the shop and of the production company of the goods, but, at the same time, while pressing the key of acceptance, also becomes the user: the user of the sw functionality that really renders the computer, for example,

10 The Bios (*Basic Input/Output system*) is a sf containing the instructions for the basic functioning of the hd and that allows a computer to open the operating system, they cede to it the control of the machine, after the verification of various parts of the hd (autodiagnostics).

an instrument of work or of amusement. The client buys a material to become a 'user' of am immaterial object or the functionality complies with various programs: the licence of use and the copyright exactly confirm this position of non possession and of non modifiability of that which I have purchased. I can alter the cd-rom, scratching it for example – that is allowed –, but I cannot modify the program if it is not within the parameters of the foreseen personalisation already in the program itself. In that way if I buy a book I own material goods, the paper and the ink, but not that which it represents as meaning the fact that ink is arranged has importance: it is necessary to quote authors and publishers, it is necessary to comply with the copyright. The common aspect is the fact that the cd-rom, as much as the paper and the ink, have a value completely laughable in comparison to that which they are transmitting. Therefore it seems one is faced with the logical implication that one foresees to being clients to buy the right to be users.

The idea of the user therefore is born when the separation between man and his competences and the machine an its competences have been established: from hw to sw clearly you pass to the alphanumeric interface (for example Dos) to get to the GUI, to the browser of the WWW, to network problems, to the wearable and, in the future, to chips inserted into the body. In all the successive passages at first, the relationship man-machine is always mediated (interfaced) from a sw that transforms the 'owner' into the user. The interface thus seems a crucial point: as long as there is the need for learning, inconsistency, unseen processes etc. the user will exist, in a specific way, a user to follow, to control, to trust etc.

The interface, like complexity, maybe will disappear when it becomes an external protection of our body. Computers, if they are still called in that way, inserted into our bodies will no longer be extensions of but will become introjections of our bodies.

At the moment the relative simplification of the interface seems to be in vain due to the accumulation of problems that the new sw, in the society of the net, prose. Therefore the simple frequenting of the web poses new problems. In terms of sw for the personal and company production, but also simply for amusement there is added, in fact, to programs which are always more integrated and interactive among themselves, things even more complex, more interactive with other machines of other users, as far as they are concerned, for example, the distance collaborative work and the sharing of data or flies. As far as the web is concerned you can signal the complexity of the procedure for the benefit of the possibility offered by the net (problems connected to the configuration of the links, of the *plug in*, etc) and in an urgent way, the problems of security. The society of the net has cut into the user in a very serious way with the problems of security, imposing from one side a major dependence on the sw which is always more complex and demands from the visions of complicated unity (knowledge of IP addresses, of domains, of compatibility, configuration of protocols, of communication etc); on the other increasingly more extensive competence connected to the system of security (proxy, firewall, antivirus, antispam, parental control, etc) that also interact with the programs installed on the machine.

The requests *of* the user grow, as well as those *to* the user. In the society of the net at least two or three types of competences are necessary: one for the interface and for the processes connected with the web browser single interface for various actions and various services usable via the internet, and the other is connected to the problems of security that for the individual are becoming a collective problem.

Thus an increase in the competences of also the basic user is requested. At first we can track down an often-implied knowledge (underlying mechanisms to some functions linked to surfing and generally connected to a practice reasonably consistent to the web with the use of e-commerce, e-learning, etc) and an explicit knowledge (relative to the simpler functions of surfing). In the second all the more technical knowledge is relegated, linked to the diffusion of not only viruses, but also of all those sw and informatic agents that operate in a hidden way behind the pages that we see via the browser and that are known as *spyware, activeX,* etc.[11], that take information from our computer, unknown to the user, sending it to a server that stores it on a *database.*

11 There little programs that are downloaded by visitors to the web, often unknown, and that collect information about the user and to send it to a server that memorises it.

We find ourselves confronting an increase of requests based more on the logic of the machine and the programs to pass, then, to one of the connections among the programs to come, today, and increasingly in the future, to the logic of sharing hw and data in workgroups, of general resources, until one arrives at the hidden objects in the fruition of single web pages and based on *cookies*, *java* programs, etc.

According to this vision the user is not only destined to disappear but will be increasingly dependent on the products and supply of services. He will also have to increase his competences in such a way so as to become not only a 'skilled user' of programs but also of relationships and connections between them and the machine. The user is no more the 'expert' of single functions but rather an 'integrated subject' active and capable of choice, that creates correlations among the elements (functions among programs, relation of contact with the interface, agents of communication, etc.) A user, therefore, that needs the skill relative to elements that can be human, artificial and mixed.

Such abilities can reshape the relationship between the user and the supplier of services, whoever it is, in such a way as to transform the final user, or as I have been made to feel 'terminal', into a figure that really occupies the centre of a network that is principally economic, tending from the start to be indebted to the considerations of the cultural peculiarities of the user.

A kind of analogy is drawn with the user of mass media. The progression of the concept of user seems thus to follow the evolution as that of the concept of the beneficiary in communication via mass media has had: from the idea of a passive audience to that of an active audience that actively and constantly negotiates the news. In the world of the PC this is the equivalent of saying that the user, including the basic user, contributes to the creation of his informatic *brainframe*[12] with his knowledge, his needs that hide in his culture, the unity of his known procedure, the informatic socialisation and his will to turn the software to his needs that will put him in a position to benefit, to the full, from his role of user. At the same time proposals for a bill of 'rights for the user' could really become fashionable as a powerful in-

12 Cf. D. de Kerckhove, *Brainframes. Mente, tecnologia e mercato*, Baskerville, Bologna 1993.

strument different to the rights of the consumer considering the specificity and importance of the field under consideration.

Bibliography

M. Castells, *Galassia Internet*, Feltrinelli, Milano 2002.
M. Castells, *La nascita della società in rete*, Università Bocconi Editore, Milano 2002.
L. Del Tutto, ' "Utente", "utenza" e forme europee comparabili', in M. Negrotti (ed.), *Homo utens*, Guerini e Associati, Milano 2004.
F. Fileni, *Analogico e digitale*, Goliardiche Edizioni, Trieste 2003.
D. de Kerckhove, *Brainframes. Mente, tecnologia e mercato*, Baskerville, Bologna 1993.
J. Nielsen, *Homepage usabilità*, Apogeo, Milano 2003.
J. Nielsen, *Web usabilità*, Apogeo, Milano 2000.
R. Scalisi, *Users*, Guerini e Associati, Milano 2001, p. 145.
M Visciola, *Usabilità dei siti web*, Apogeo, Milano 2000.
A. Wilden, voce *Comunicazione*, Enciclopedia Einaudi, 3rd vol.

Franco Fileni, Dip. di Scienze dell'Uomo, Università di Trieste

Giuseppe Lanzavecchia

A conceivable model for user: fields and attractors

Abstract:

A model of users requires coexistence between users and suppliers; each side stimulates the other, and the suppliers compete with each other. This is typically a field containing more or less stable attractors (the suppliers) and characterized by the presence of other operators – laws, policies, media, technology, culture, values, behaviors, and so on – which, acting as constraints, condition it. The user field consists of elements sufficiently numerous and diversified to be considered 'thermodynamic'; that is, subject to the laws of this theory and of statistical mechanics. The field evolves over time, due to changes in users, suppliers and constraints.

Keywords:
model, field theory, chaos, attractor, user, usership, supplier, causality, statistics, information, thermodynamics, physical phenomena, social phenomena, individual, society, reductionism, holism, bifurcation, constraint, change

1 Foreword

The paper suggests a model of users and usership, albeit the author thinks [Lanzavecchia, 2004[1]] that models, though useful, cannot fairly represent real situations, because they add spurious aspects and are artificial constructs representing only one facet of reality, never its entirety [Negrotti, 2002].

The proposed model needs to consider a number of users large enough to form a diversified ensemble to which the methods of statistics and thermodynamics can be applied. Causality appears in physical and social phenomena as an effect of statistics. The laws of statistics apply in all cases and conditions and the causality that characterizes the world we live in is due to the huge number of particles, or of events involved in observed phenomena – so huge as to make *certain* results that would otherwise be merely stochastic. The number of suppliers can be small (oligopoly) or single (monopoly).

Many userships are for disparate areas of interest, but even when these areas offer tangible or intangible goods that apparently have nothing to do with each other, they end up competing and thus interfering with each other, because the resources available to users are limited, as is their time. Cell phones derail business from bars and movie theaters.

Users need an ecosystem that enables information to be provided. Lastly, there are innovation-devourers, users who select innovations. An innovation cannot exist unless it is shown, in particular to users [Latour, 2004]. Well-being in our individualistic western society – marked by urgency and intensity – is increasingly characterized by excess, and the user ends up generating instability, volatile relationships and conditions, and contingent cycles.

The foregoing postulates suffice to tackle the problem of how to build a model of usership.

2 A first statistical model of users and usership. The theory of fields

Physical agents (electricity, magnetism, gravity, and so on) and to an increasing extent social and psychological agents (markets, services, empathy and hatred) can be studied by examining the physical and the conceptual regions that they influence. A field can be pictured as an ensemble of curves (more or less dense *lines of force or flow*) that represent magnitude of force, and direction exercised by the agents in the regions.

Petrus Peregrinus de Maricourt was the first to use lines of force to picture the magnetism produced by a sphere of magnetite, and thus the first to associate a field and a force. But it was Michael Faraday who conceived – with the methodology of modern science – the fields whereby a system's energy is no longer situated in its components (bodies, entities), but rather in the space in which they are immersed. Faraday's intuition changed the model of reality. In the old one, objects (or entities) were distributed in absolute space and time (as defined by Newton and Kant) and interacted along straight lines. In the new model, bodies are immersed in a space that contains and manages their properties and determines how the properties affect the bodies.

James Clerk Maxwell developed equations to formalize Faraday's ideas about electromagnetic fields, and showed that they can generate and transmit light (photons). Albert Einstein extended field theory to masses. More recently, quantum mechanics was used to quantize fields.

While Newtonian physics and thermodynamics theory use homogeneous and isotropic spaces in which events occur, field theory postulates non-linear spaces where the lines of force are not necessarily straight; as a rule, they are curved.

The field model has obvious philosophical and ideological postulates. Faraday, for instance – like his teacher Humphry Davy and his friend Christian Oersted – was deeply religious, and his philosophy was very much like the 'Naturphilosophie' of the Romantics, especially Schelling. In their eyes, the field model overturned the one that French science had worked out during the Revolution – the model of interaction among bodies, which they considered atheist. Faraday and Davy started from the concept, formulated by Ruggero Giuseppe Boscovich, that the natural basis of the elements was not matter; it consisted of mathematical points surrounded by *fields* of forces, positive or negative, attractive or repulsive. This influence of philosophy on the scientific concept of fields continues in our own day [Lange, 2003].

Field theory works in many very different cases, from magnetism to usership; according to Johnjoe McFadden, our conscious mind might be an electromagnetic field that permeates space; Kurt Lewin has tried to frame the social sciences in field theory. Mauro Dorato has shown how many scientific theories – often expressed in mathematical models – fit a wide variety of physical and social phenomena:

there would seem to be some sort of universal software permeating the laws and events of nature.

Can we use a model like field theory without going into the physical, behavioral and psychological details of the phenomenon we want to study and explain? If we do, our construct will be able to provide only a very partial description of reality. The field is in fact generated by individuals interacting necessarily with each other, but it is also determined by the rules that govern community life and the ideologies that underlie social relationships; it is a synthetic tool that incorporates individuals and events in a statistical conception of the social system.

Individuals cannot be eliminated from the scheme; indeed, our western society places them at the center of all events. From the ethical standpoint, the most important aspect is responsibility. For upholders of the individual versus society, responsibility is individual, but there is also a long tradition that ascribes responsibility to society.

Individuals or field? If a system comprises many individuals, they cannot be considered one by one; the ensemble of actions and events must be viewed statistically. In this case, however, the field model incorporates causes and effects, actions and consequences. In mathematical terms, the two descriptions – of the ensemble of individuals and of the field that incorporates them – can be formulated dually and correlated via a Fourier transformation.

The ideological antithesis remains, however. It is the same one found in every scientific description, because a choice must be made between, for example, a reductivist or Galilean approach (minimum number of essential parameters, as already required by Ockham's razor centuries before Galileo) and a holistic or Batesonian approach (the whole universe). The former ends up idealizing experiences and results, the latter is simply unworkable. In either case (without going into the many other aspects), one can only consider a finite and small number of parameters [Lanzavecchia, 2004[2] and Longo, 2004].

3 Deterministic chaos

Field theory can describe the characteristics of a social system quite well, in more or less the same way that a snapshot describes a scene; it has the virtue of reducing the weight of the individualistic aspects (though they are the basis for the very existence of the field) and thereby augmenting the role that the field itself plays in determining the behavior of the individuals who populate it and are subject to its forces. The influence of the field can be quite strong; suffice it to think of the effect of fashion news on consumer choices, or the desire to imitate that in the last few years has driven practically the whole population of Italy to buy cell phones. It would seem that the characteristics of the user-supplier system are no longer to be found in the actors or their actions; rather, they are an integral part of the field in which the actors operate. Statistically speaking, their behavior and actions are determined by the field.

But although field theory produces fine *snapshots*, it cannot do a good job of describing the chaotic, convulsive changes in situations and characteristics that are typical of our times. Deterministic chaos theory fits the bill better, for it was conceived to describe systems that contain elements of disorder and evolve as new actors (suppliers and users) appear and others disappear. Such systems are dynamic, non-linear and can present quasi-periodic motions.

How the system behaves depends on its initial state and on the values of the parameters. Two identical systems that differ only, and very slightly, in their initial conditions may end up very different after a finite period of time. Changes in the weather are an example of this sensitivity to initial conditions.

A great variety of chaotic phenomena can be studied by applying the theory of deterministic chaos and imagining that they occur in spaces that are congenial to their characteristics.

This is certainly not the place to describe chaos theory, even in very general terms. Here I would simply note that the dynamic models considered up to the end of the 19th century were linear; they represented force-deplacement as straight lines in a one-dimensional space. A typical case is that of a spring hanging at one end from a rigid support, with a weight attached to the other end. If we pull on the spring and then release it, the ball bounces up and down, preserving its energy. But if the system is immersed in a viscous environment, the system's energy is absorbed by the environment until the ball stops at the lowest possible value. What we have here is a *dissipative* system; the lowest point where the ball stops is called a *strange attractor*, and the straight line along which the weight moved is its *attraction basin*.

The system I've described is a very simple one, but it's not hard to think of more complex systems – in three-dimensional physical spaces or many-dimensional phase spaces, for example – and of different attractors, due not only to weights but also to magnetic poles, electrical charges, or offerings of useful and desirable goods and services. And we can also imagine other attraction basins, such as the domestic or world market, the communications or fashions market, and so on.

Attractors are pits of potential in which elements of the system can be captured. In the case of linear systems and two-dimensional space, the pit is parabola-shaped, while in three-dimensional space it is paraboloid. But if the system is not linear, the pit's shape is more complex. The slopes are not uniform; they may be steeper here and flatter there. There may be plateaus that slow the pit's capacity to attract elements of the system, and surrounding hills that make access and exit difficult and thereby unstably increase its capturing capacity. The pit may also have other features that reflect the presence of operators, constraints and particular situations.

The space in which our system is immersed may contain multiple attractors with very different features that act on the system independently or in direct or indirect competition with each other. As the system *moves*, it may come upon a fork (said *bifurcation*): a situation in which it is confronted with two attractors – that is, two alternatives.

In reality, our system is not isolated from the rest of the world. The world continuously stresses the system with disruptions that *stir* it and make all measurements intended to locate it imprecise. The theory underlying this non-linear dynamics was born with Poincaré and developed in the 20th century with contributions from many scientists and from the powerful computing methods offered by information technology. René Thom first used the word *catastrophe* to signify an abrupt change due to attractors that change the spatial conditions in which a system operates, thereby nullifying the evolutionary history of the past.

Chaos theory uses the theory of fields containing attractors that determine the nature and configuration of the space in which a system operates. Attractors may appear and gather strength, or wane and disappear. The system may become more and more complex by choosing, when it comes to a bifurcation, the path most interesting for its evolution.

4 The structure of the user field

The user field is characterized by a multitude of other *operators*: laws and regulations, culture, values and behaviors, the political system, the media, technology. These operators act as constraints, but also as elements that modify the field, favoring or penalizing or even destroying the attractors, or stimulating the growth of new ones. Even if the ensemble of constraints limits many of its possibilities, though favoring others, the field is *free* enough to be considered *thermodynamic*; in other words, it is subject to the laws of this theory and of statistical mechanics.

One of the aspects that guarantee that a system is thermodynamic is a lack of explicit intelligence; that is, the system is not consciously steered by any of its components toward specific objectives. An analysis of the stock market indicates that its operators act haphazardly and that economic fluctuations seem to be the result of random decisions [Ball, 2003]. If the 'geniuses' of finance act on average with *zero intelligence*, it is safe to say that users do the same. More in general, we can say that politics does not think, because society as a whole does not do so [Canto-Sperber, 2004]. This does not mean that there is no rationality in the actions of users and investors, society and politician, and that it cannot be identified ex-post and intuited ex-ante.

A tribal structure bound by behaviors that are strongly predetermined by rules, values and taboos cannot accept changes due to factors that deeply disrupt these behaviors. The tribe either eliminates the changes or is destroyed by them. Only free and open societies are assuredly thermodynamic.

The user field evolves over time as the result of changes in suppliers' offerings, in user behavior and needs, and in the constraints. The suppliers are and remain attractors if they respond to user needs, and continue to do so.

The proposed model can be enhanced by separating the role and representation of suppliers and constraints, representing suppliers as attractors and incorporating constraints in the space, as the 19th century physicist Heinrich Rudolph Hertz did. Hertz worked out a mechanics that does not use the concept of force, but translates forces into constraints.

The user moves in a space containing constraints and stimuli that reflect the existence of laws and regulations, traditions and values, the thrust of fashions and innovations, and many other aspects that condition the everyday and long-term behavior of individuals and groups.

5 Can the field-chaos model represent users' behavior and characteristics?

The author has never believed in the existence of fields and thinks that physical phenomena are due solely to direct interaction among elements without any mediation, such as that of fields.

In this conception of events one can take into consideration only isolated phenomena. Phenomena involving a great number of elements require to use collectivist tools such as Maxwell's and Boltzmann's statistics, in the case of classical physics. For these statistics, the H theorem applies and according to it, a system, in whatever state it is in, will tend to reach the ideal configuration of equilibrium, provided that its total energy is not perfectly determined. For elementary particles, we must use the Fermi-Dirac or Bose-Einstein statistics.

The same reasoning should apply in sociology, in particular to usership. A sociological phenomenon that involves a great number of elements but can be (arbitrarily) considered as independent of others should be studied with rigorous statistical methods [Negrotti, 2002[2]]. If we intend to consider multiple groups of interacting phenomena, however, we need to find other solutions: in the case of usership, I suggest the theory of fields, deterministic chaos and space that incorporates constraints.

Field theory was conceived to consider phenomena in equilibrium; chaos theory applies even in conditions of non-equlibrium and in this case we must use Onsager or Prigogine thermodynamics.

Constraints, too, change over time. Laws and regulations are intended to be static – in equilibrium – in the hope that they will last forever. However, the convulsive changes taking place in society today are such that laws and regulations are often outdated even before they are enacted, or at any rate very soon after [Lanzavecchia, 2002]; in other words, they are generally out of equilibrium. The media continue to send out disparate messages with no substantial logical structure and they are chaotic. Our democratic political systems are certainly unstable – not in equilibrium – while many authoritarian systems can be said (with some approximation) to be in equilibrium. Culture, values, behavior and fashion, in that ascending order, are unbalanced if not fibrillating. Technology (and the science) is subversive, in the sense that it dramatically changes the conditions in which the constraints, operators and users act. Changes are not brought about by technological breakthroughs but even more are the result of incremental innovations that change every situation, product, activity and concept day by day, making everything quickly obsolete [Schrage, 2004]. According to Emanuele Severino, 'technics will defeat economics and politics'.

The proposed model is useful for examining and understanding the contrast between conservative and innovative trends in political, economic, social and entertainment offerings, and users. It must remain qualitative because it cannot be translated into mathematical equations, primarily because of the lack of data, but also because of the size and complexity that such equations would have. In some simpler cases, however, it would be possible to work out quantitative descriptions.

Bibliography

P. Ball, 'Stock market traders show signs of zero intelligence. Economic fluctuations appear to result from random decisions', in *Nature Science*, 24 September 2003.

M. Canto-Sperber, 'La politique peut-elle penser?', in *Le Monde*, 19 March 2004.

M. Lange, *An Introduction to the Philosophy of Physics: Locality, Fields, Energy, and Mass*, Blackwell, London 2003.

G. Lanzavecchia, 'Evoluzione degli ecosistemi urbani', in *Ecosistemi Urbani,* Convegno, Accademia Nazionale dei Lincei, Roma 2002.

G. Lanzavecchia, 'Models in science are unnecessary but useful', in *Yearbook of the Artificial, Vol. 2, Models in Contemporary Sciences*, M. Negrotti (ed.), Peter Lang, Bern 2004.

G. Lanzavecchia, 'Riduzionismo e olismo, tra Galileo e Bateson', in *La Nuova Scienza, Vol. 5, Al di là dello specchio*, U. Colombo and G. Lanzavecchia (eds.), Libri Scheiwiller, Milano 2004.

B. Latour, 'Nessuna innovazione senza rappresentanza', in *Il Giornale dell'Ingegnere*, n 1, 15 January 2004.

G. O. Longo, 'Una nuova scienza?', in *La Nuova Scienza, Vol. 5, Al di là dello specchio*, U. Colombo and G. Lanzavecchia (eds.), Libri Scheiwiller, Milano 2004.

M. Negrotti, *Naturoids. On the Nature of the Artificial*, World Scientific Publishing Co., Singapore 2002.

M. Negrotti, 'Patologie del Novecento', in G. Lanzavecchia and M. Negrotti, *In difesa della scienza*, Libri Scheiwiller, Milano 2002.

M. Schrage, 'Disruptive incremental innovation', in *Technology Review,* March 2004.

Giuseppe Lanzavecchia, LCA – Laboratory for the Culture of the Artificial, University of Urbino

Julian Locke

On-line machine-based natural-language translation: a new approach

Abstract:

An example of an artificial system that has become increasingly common in recent years is the on-line translation service, intended to produce translation of texts from one human language to another. However, current-day on-line translation services continue to produce unsatisfactory results. In this short essay, an attempt is made to outline a means by which such services may be improved, partly by reversing a tendency towards oversimplification, which has led to a marginalization of the user and a subsequent loss of overall utility, and partly by allowing users to contribute to the continuing expansion and enhancement of the services, in order that they may become ever more useful and trustworthy tools.

Keywords:
Internet, on-line translation services, learning ability, over-simplification, user input

There are now numerous on-line translation services to be found on the Internet. However, anyone who has ever tried to use one of these services will almost certainly have been disappointed, or at best rather amused. Indeed, anyone who has tried a number of such services will know that the quality of translation they provide is far from satisfactory. While we have seen huge improvements in search engines and other on-line services, translation services continue to fall behind.

Of course, no one could reasonably expect a machine-based translator to rival a bilingual human translator, but it does not seem unreasonable to expect computer-based translation to be of genuine *use* to the *user*. As things stand, however, it is far from clear that machine-based translation services provide a genuine service.

Below are presented three examples of translations performed, from English into Italian, by web-site machine-based translation services. I used the services to translate the same short piece of text, in order to compare their results[1].

The original text reads:

> The shared meaning of the word 'model' has changed a great deal during the last three centuries, shifting from something real to be reproduced to something artificial that tries to reproduce something real. This shift is probably due to the gradual discovery of the role of the mind in trying to reproduce natural objects or processes in different ways. In fact, today it is widely recognized that, when we describe the world, we are actually describing our models of it. This point has even been emphasized to maintain, as constructivists do, that reality does not exist at all.

Translation by WorldLingo.com:

> Il significato comune della parola 'modello' è cambiato molto durante gli ultimi tre secoli, spostando da qualche cosa di reale da riprodurre a qualche cosa di quel artificiale prove riprodurre qualche cosa di reale. Questo spostamento è probabilmente dovuto la scoperta graduale del ruolo della mente nel provare a riprodurre gli oggetti o i processi naturali nei sensi differenti. Infatti, oggi ampiamente è riconosciuto che, quando descriviamo il mondo, realmente stiamo descrivendogli i nostri modelli. Questo punto persino è stato dato risalto a per effettuare, come i constructivists, quella realtà non esiste affatto.

Translation by Freetranslation.com:

> Il significato diviso del 'modello' di parola ha cambiato un grand'affare durante gli ultimi tre secolo, spostando da qualcosa di reale di essere riprodotto a qualcosa di artificiale che tenta di qualcosa di reale di riprodurre. Questo turno è probabilmente dovuto alla scoperta graduale del ruolo della mente in tentare per riprodurre gli oggetti naturali o i processi nelle maniere diverse. Infatti,

1 The text in question is taken from the Yearbook of the Artificial, Vol. II, edited by Professor Massimo Negrotti of the University of Urbino, Italy, and is the first paragraph of the abstract of the essay 'Models Thinking', written by Professor Negrotti himself (Y.O.A. II, p.23).

> oggi ciò è largamente riconosciuto, quando descriviamo il mondo, descriviamo effettivamente i nostri modelli di esso. Questo punto è stato anche accentuato per mantenere, come il constructivists fa, quella realtà non esiste a tutto.

Translation by Systran.com:

> Il significato comune della parola 'modello' è cambiato molto durante gli ultimi tre secoli, mobili da qualche cosa di reale per essere riprodotto a qualche cosa di quel artificiale prove riprodurre qualche cosa di reale. Questo spostamento è probabilmente dovuto la scoperta graduale del ruolo della mente nel provare a riprodurre gli oggetti o i processi naturali nei sensi differenti. Nel fatto, oggi ampiamente è riconosciuto che, quando descriviamo il mondo, realmente stiamo descrivendogli i nostri modelli. Questo punto persino è stato dato risalto a per effettuare, come i constructivists, che la realtà non esiste affatto.

Now, these translations are not entirely without merit. They could certainly be of use to an Italian in interpreting the original text, in order to gain a basic understanding of it. Yet each of them shows the shortcomings of current-day on-line translation technology. Each contains numerous errors, and even the best of them would need considerable editing to produce a satisfactory Italian translation of the original text – and this, of course, is one of the major difficulties with on-line translation: if it is not done well, it is arguable that it might as well not be done at all. One might well be better off with a good dictionary and some grammar books.

It is noticeable, in particular, that none of the three translation services has recognized the word 'constructivists', yet each of them has simply inserted the original English word into the supposedly translated text, giving no indication that it has not actually been translated. In fact, this behaviour – of not indicating unrecognized words – is common to *all* of the on-line translation services I have found on the Internet. Yet this is not a difficult problem to overcome, and it is hard to understand why it has not been dealt with. All that is needed is a way of marking, in the output text, any words that have not been recognized by the translation program. They could, for example, be underlined, or written in Italics, or perhaps shown in a different colour.

Having identified an unknown word, however, the *user* will still have to translate it, and, as we have already established, the on-line translator will not be of any help. Instead, the user will need to resort to another source, such as an English-Italian dictionary. Moreover, having successfully translated the word, he or she will have no way of updating the translation service to include the word, and the next time he or she, or anyone else for that matter, submits a text containing that same word, the on-line translator will once again fail to provide an adequate translation of it.

This brings us to one of the principal problems with on-line translation services. They fail to learn. Generally, the user is asked to type in some text, or to insert it from another source, choose the source language and desired output language, and then press the 'Translate' button. Then, when the translation is returned, the user has no choice but to accept it as it is. This inability to learn would not be a serious problem if on-line translation services were good at what they do, but, as yet, they certainly are not. (That is not to say that such services never improve. Every so often the translation engine is replaced, resulting in better translations and therefore greater utility. Nevertheless, the sense of progress is distinctly limited.)

It might be argued, of course, that the inclusion of more advanced facilities on translation service web-sites would add to the complexity of the service, thereby confusing, or even alienating certain would-be users. Perhaps in an effort to increase the accessibility and utility of such services to an ever-increasing proportion of the public, the developers have tended to maximize the ease of use, even if at the expense of more advanced, and potentially more useful, facilities. Yet if the true expense of omitting such facilities is an unsatisfactory (and, at times, all but useless) translation, the appeal to simplicity seems misguided.

The problem of oversimplification is not limited, of course, to online translation services alone. Rather, it is indicative of a more general phenomenon that we have been witnessing for some time. Since the advent of computers, attempts have been made to provide users with more and more elaborate facilities; yet, at the same time, every effort has also been made to provide those same users with more and more user-friendly interfaces. That is to say, the user has been given ever-more-powerful tools with ever-simpler controls. My question, however, is: has this combined process gone too far?

We have certainly come a long way from the early command-line-based user interface, which required the user to have considerable knowledge of the workings of the operating system in order to achieve the simplest of tasks. Now, through various stages of development, we have come to live in a point-and-click, drag-and-drop age, in which anyone with a reasonable intelligence and a bit of courage can perform tasks that little more than a decade ago would have required considerable expertise.

Yet with each new development, the typical user is becoming an increasingly passive participant in the partnership between man and computer, interacting – particularly with the most widely dispersed computer applications – in a very restricted way. What is lacking, I suggest, is a *trust* in the users of computers and information technology (although I shall later qualify that statement). At the moment, it seems, the trust is in one direction only. The user must trust the computer, and have faith that the programmers have produced properly working code; but the programmers, in general, do not return the trust. Indeed, they devote much time and effort to preventing the user from pressing the wrong button or inputting inappropriate values. It is as if they see the user as an unfortunate but unavoidable participant in a system that would be far cleaner and more efficient without any human input at all.

While this is perfectly understandable on one level – as erroneous input, whether deliberate or accidental, could certainly give rise to problems in many circumstances – I suggest that it is also potentially counterproductive and limiting. Specifically, it is failing to make use of a most valuable resource – namely, the knowledge and intelligence of the millions of ordinary people using computers throughout the world.

Machine-based translation services, in particular, constitute systems that might well benefit from additional user input. My proposal is to provide certain users with the means to furnish such services with additional vocabulary, in order that they may be improved with continuing use. One of the difficulties, however, with allowing users to play an active part in improving an on-line service, or even a stand-alone computer application, is that user input cannot, in general, be considered authoritative, and has the potential to damage or to undermine the usefulness of the service. In other words, the programmers are *right* to be cautious when considering the human element in computer systems. Nevertheless, it seems blinkered and short-sighted to restrict process merely to retain stability and to avoid risk.

I wish to promote the idea of a particular form of distributed computing. At this moment, there are numerous instances of distributed computing taking place around the world. One commonly used system is the 'farming out' of blocks of data to hundreds – or even hundreds of thousands – of computers connected to the global network. Users are generally provided with a program designed to run in the background when the system is not busy. Two of the best-known examples of these are the SETI project – the Search for Extraterrestrial Intelligence – and the GIMPS project – the Great Internet Mersenne Prime Search[2].

These projects, however, and others like them, involve minimal interaction with the individual computer-user. They generally run as screen-savers, and the only decisions the user ever need make are when to activate the screen-saver and when to connect to the web-site to submit the results and obtain more data to work on. Moreover, distributed computing in this sense is not suitable for all large-scale computing projects involving many computers connected to a network, and it is difficult to see, in particular, how an on-line translation service could make use of such a system.

2 The SETI project is in fact the largest computing project ever undertaken, now totalling over two million years of computing time. Yet the true power of distributed computing is perhaps demonstrated more strongly by the GIMPS project, which was set up to look for very large prime numbers. For mathematical reasons, only numbers of the form 2n–1, known as Mersenne numbers, are tested. These numbers yield to particularly efficient algorithms for testing primality. Mersenne Primes become increasingly rare as n increases, and to date only 42 have been discovered, the last 8 of which were found by GIMPS, which has now been running for over 8 years. In February 2005, the 42nd Mersenne prime, $2^{25.964.951}–1$, was found. Containing over 15 million decimal digits, it took more than 50 days to be checked for primality on one of the millions of computers engaged in the search throughout the world. The discovery of this number would have been very much later in coming were it not for the use of distributed computing, which creates, in effect, one vast parallel processor that is a match for even the fastest supercomputers.

Nevertheless, a sort of distributed data-input might, I think, be implemented. My proposal, as I have suggested, is to allow certain (suitably chosen) users to play a part in improving and updating translation services. A given on-line service could provide such users with a form with which they could submit additional words or phrases to the lexicon. In this way, they would be able to update the translation database, or, more probably, to add to a supplementary database. The structure of this form is, of course, all-important. It must be able to provide the user with a sufficiently rich set of features to allow him or her to specify with precision the nature of the words or phrases to be added, in order that the additions may be properly integrated. The creation of a suitable interface would be a learning process in itself, and would very likely require numerous to accommodate previously unconsidered exigencies.

If active contributors could be chosen according to certain criteria, and a process of double-checking implemented, whereby submissions made by one user could be verified, or perhaps disputed, by others, then the risk of erroneous input would be minimized. Prospective contributors would first have to register and be subjected to a vetting process, after which, if accepted, they would be allowed password-protected access to the submission form. For stand-alone software, this sort of extended interaction would generally be unproblematic, of course, as other users would not normally be affected by changes that a given user made to his or her own computer. With an on-line service, however, changes in the service would have the potential to affect all those who use it.

I have written a very simple program to demonstrate the sort of participation I imagine[3]. The program presents the user with the sort of interface that one might find on a typical web-site providing a machine-based translation service. As is normally the case for an on-line translation service, the user may type in some text in the source language, then click on 'Go' to see the translation appear immediately below the source text. My simple program is intended only as a demonstration, and has only a very limited vocabulary. Nevertheless,

3 The demonstration program may be obtained by request, by writing an e-mail to one of the addresses given at the end of this paper.

where it differs from most on-line translation services is in its ability to provide the user with the means to add new words. Thus, for example, when translating from English to Italian, if the English word *constructivists* has not been recognized initially, the user may choose to access a form which will allow him or her to add the word, and to provide its Italian equivalent, *costruttivisti*. The user may also provide extra information, such as gender, plurality and grammatical category, which will help the translation program to produce a more accurate translation. Thus, for example, *costruttivisti* is a plural masculine noun in Italian, corresponding to a plural neutral noun in English. The extra information would allow the translator to choose the appropriate form of any articles or adjectives that accompany the noun. This is particularly important in Italian, as articles and adjectives change their form according to the gender and plurality of the nouns to which they apply, whereas in English they are invariable.

Another consideration is whether an adjective is prepositive or postpositive – i.e., whether it comes before or after the noun it qualifies. Most, but not all, adjectives in English are prepositive, while most, but again not all, adjectives in Italian are postpositive. To further complicate matters, some adjectives may be placed either before or after the noun they qualify, but often the word-order will depend on the context, and what sounds right in some circumstances may sound odd, or pompous, or just plain wrong, in others.

One of the more difficult problems to deal with when translating between English and Italian is that of verbs. Italian verbs are highly inflected, giving rise to about ninety different forms of each verb, declined using the verbs *avere* (to have) and *essere* (to be). The verb-structure of English is very much simpler, with almost no inflection, and different tenses and moods are formed by the use of various auxiliary verbs. On the other hand, English includes more than three thousand phrasal verbs, of which many are in common use. Phrasal verbs cause particular difficulties for machine-based translators, not least because the words constituting the verb may often be separated by other words. Moreover, many phrasal-verb constructions have two or more meanings, generally distinguished by the context. Perhaps more difficult still are the idiomatic expressions and the proverbs of a given language, many of which have equivalents in other languages, but not always word-for-word equivalents. Thus, for example, 'deaf as a post' or 'stone-deaf' would be equivalent to '*sordo come una campana*' (literally 'deaf as a bell') in Italian, and '*sourd comme un pot*' (literally 'deaf as a pot') in French. Similarly, the English proverb 'A bird in the hand is worth two in the bush' is approximately equivalent to the Italian proverb '*Meglio un uovo oggi che una gallina domani*' (literally 'Better an egg today than a hen tomorrow')[4].

These are just some of the considerations that must be taken into account when attempting to translate effectively between languages. It is not difficult to predict that creating an on-line translation service with an ability to learn will be a tricky business, requiring considerable expertise and foresight. However, with the existence of numerous on-line services already providing machine-based translations, the basic structure is, I suggest, already in place. All that is needed is a way of gathering additional information from users, and then assimilating it into the existing structure. Then, with the application of due caution, the currently passive user may be allowed to play a greater role in shaping the service that he or she seeks to receive, and, in doing so, will be able to hope for a service that will continue to improve with every contribution made.

Julian Locke, LCA – Laboratory for the Culture of the Artificial, University of Urbino, *jlocke@onetel.com* or *english.lessons@libero.it*

4 An alternative version is '*Meglio l'uovo oggi che la gallina domani*' (literally 'Better the egg today than the hen tomorrow'). There is also another Italian proverb with much the same sense, which more closely resembles the English version – '*Meglio un fringuello in gabbia che un tordo in frasca*', which may be translated roughly as 'Better a chaffinch in the cage than a songthrush on the branch'. The French equivalent is '*Un tiens vaut mieux que deux tu l'auras*', which translates, according to www.freetranslation.com, as 'A holds is more better than two you will have it', thus illustrating my point all too clearly.

Riccardo Manzotti and Vincenzo Tagliasco

Towards an artificial user: the 'what' problem for an architecture capable of developing new goals

Abstract:

A user is a system capable of creating and pursuing individual goals. Is it possible to design and implement an artificial user? Traditional artificial systems focus on how achieving a given goal. Most learning algorithms look for an optimal solution of a problem, given a set of optimization criteria and a goal (or a set of goals). However, real agents and real users have to develop new goals in order to cope with their environment. They must find 'what' they want to achieve and not only 'how'. The development of completely new goals on the basis of the interaction with the environment is here defined the 'what' problem. In this paper we will try to define it and we will propose an architecture capable of addressing it. Such an architecture is proposed as the foundation for an artificial user.

Keywords:
ontogeny, phylogeny, motivations, robot, development, goal

1 What and how

What is a user? A user is a system with its own goals actively trying to purse them by means of tools or interacting with other systems. How does a cognitive system, like that of human beings, produce its own goals? And why do current artificial systems lack this capability?

Current implementations of artificial systems focus on the implementation of intelligent algorithms to implement optimal performances to achieve a fixed goal. Human users show a different approach. They not only achieve an optimal performance with respect to a given set of hardwired goals; they develop unpredictable and unexpected new goals. The new goals are totally or partially dependent on a limited set of hardwired goals fixed at design time; they are the result of the interaction with the environment and can overcome the original ones.

The objective of this paper is twofold: to outline the 'what' problem and to describe an architecture capable of addressing it.

Artificial systems are designed with an already fixed set of goals that has to be achieved. Designers focus their efforts to find 'how' these goals can be achieved.

Learning is defined as a modification in our behaviour driven by a goal. Various learning paradigms focus on this modification of the behaviour: Supervised and Unsupervised Learning and Reinforcement Learning are valid examples. For instance, Sutton and Barton claim that 'reinforcement learning is learning what to do – how to map situations to actions – so as to maximize a numerical reward signal' [Sutton and Barto 1998]: the goal is expressed by a reward signal which is *a priori* defined. They claim that 'the basic idea is simply to capture the most important aspects of the real problem facing a learning agent interacting with its environment to achieve a goal. [...] All reinforcement learning agents have explicit goals. These explicit goals are fixed at design time. We could say that Reinforcement Learning deals with situation in which the agent seeks to achieve a goal despite uncertainty about its environment.

We want to address, in this paper, the process by which it is obtained 'what' (the goal) the agent has to achieve. This is the 'what' problem. Using the Reinforcement Learning terminology we could say that the 'what' problem is equivalent to looking for new reward functions. In Reinforcement Learning systems 'the reward function must necessarily be unalterable by the agent'. On the contrary many biological systems are capable of developing partially or totally unpredictable goals. We propose to design systems that develop their own reward functions: systems capable of becoming artificial users.

A few examples.

Romeo meets Juliet at a Capuleti party. Nothing in his upbringing forced him to be in love with her. However after his first encounter, to meet her again is his most important goal. All his behavior is directed to achieve this new goal.

After a few minutes a mallard duck opens its eyes and sees, as a first large moving object, the face of Konrad Lorenz. Subsequently it tries to keep it in its field of view as much as possible. It aims at this new goal.

In the aforementioned examples, a new goal is added to the causal structure embodied by an agent. Biological agents, by means of natural selection or personal experience are able to do so. In them goals must 'self emerge' since they cannot receive their goal from the outside

To develop new goals is important since the environment cannot be completely predicted at design time. In nature there are basically two situations in which new goals are added: phylogenesis and ontogenesis. During phylogenesis a new species develops completely new goals that arise from the modification in the environment that are the product of the emergence of new ecological balances. During ontogenesis, accordingly to the degree of behavioral plasticity, each individual is capable of developing new goals from its experience. New goals can be a better specification of the original one or can be completely new. For instance imprinting in mallard ducks can be seen as simple example of specification of sub goals. The more general hardwired goal of 'looking for a mother' becomes the specific goal of looking for a given specimen of mallard duck (or Konrad Lorenz himself!).

2 Architectures for agents: a taxonomy

Not all the goals of biological systems are fixed at birth: they only possess a very limited, survival driven, built-in set of goals. As they grow, biological systems generate new goals on the basis of two separate factors: their genetic background and their past experience. Both are necessary to select a particular goal.

Behaviour-based artificial structures develop on experience and goals defined elsewhere at design time [Arkin 1999]. Goals are the result of the interaction between experience and a limited number of hardwired instincts (the ones provided by genes). In many complex biological systems, it is possible to distinguish between phylogenetic aspects and ontogenetic ones, nature versus nurture [Gould 1977; Elman, Bates et al. 2001; Ridley 2004]. Instincts are not goals. They are not the result of an ontogenetically caused motivation. They are a set of procedures to produce goals, as the imprinting procedure of birds.

What is a goal? When an agent wants an event to happen again, a new goal springs up. In this respect a goal is projected towards the future; this is only half the story. Why the agent should be willing such an event to happen again? Obviously because of something that happened in the past. In this respect a goal is a projection of the past. Goals are causal structures that link the past with the future. Because of this, they are the ideal candidate as a building block for an agent.

Is it possible to implement instincts and goals in an artificial system?

We propose a taxonomy of architectures: a fixed control architecture, a learning architecture and a goal generating architecture (Fig. 1). In the first case, the system has no capability of modifying *how* it does *what* it does. There is a input-output mapping module, which takes the input signal and produces the output on the basis of some a priori hard-wired module. Examples of this structure are simple control devices or machine automata. In the second case, the system is capable of modifying its behavior to fulfill some a priori target. The system is capable of modifying *how* it behaves. Examples of this structure are reinforcement learning or supervised learning artificial neural networks. In the third case, the system is capable of modifying not only *how* it does what it does, but also to define *what* it does. The goal module sets the goals that have to be pursued by the learning module.

On the basis of these three levels, three kinds of architectures are proposed: fixed control architectures, learning architectures, goal generating architectures.

2.1 Fixed control architecture (input-output)

In this case, the causal structure of the system is fixed. There is no ontogenesis whatsoever. Notwithstanding the behavioral complexity of the system, everything happens because it has been previously coded.

A mechanical device and a complex software agent are not different in this respect: both are pre-programmed in what they must achieve and how they must achieve it. Nothing in their structure is caused by their experiences. Suitable examples of this category are Tolam's artificial sow bug, Braitenberg's thinking vehicles [Braitenberg 1984], Brooks artificial insects and recent entertainment robots like Sony AIBO and Honda's humanoid ASIMO (2002).

2.2 Architecture for learning ('how')

A different level of structural dependency on the environment is provided by the architectures that can learn *how* to perform a task. Behaviour-based robots can be classified in this category. Systems based on artificial neural networks are well-known examples of this kind of architecture. These systems determine how to get a given result once they have been provided with a specific goal. The goal can be given either as a series of examples of correct behavior (supervised learning) or as a simple evaluation of the global performance of the system (reinforcement learning) [Sutton and Barto 1998]. In both cases some kind of learning is applied. These systems lack the capability of creating new goals. There are several examples of this kind of learning agent: Babybot at LIRA-Lab [Metta, Manzotti et al. 2000], Cog at MIT [Brooks, Breazeal et al. 1999].

2.3 Goal generating architecture ('what')

A system that learns both *how* to perform a given task and *what* task must be performed, corresponds to an ontogenetic architecture. This is the case for most, if not all, mammals; it is true for primates and for human beings. They are systems capable of developing new goals that do not belong to their genetic background. In the field of artificial systems there has been a series of attempts to address this problem [Fukushima, Okada et al. 1994; Weng 1996] as well as attempts to locate similar structures in the cortical architecture of humans [Togawa and Otsuka 2000]. For their development, these systems depend more on the environment than the previous two categories. A system belonging to the first category does not depend on the environment for what it does or for how it does what it does.

3 The what problem

An ideal autonomous system must be able to develop in a completely unknown environment. In order to do that, it must be provided with sensory and motor capabilities. It is reasonable to assume that the sensory and motor capabilities delimit a more limited world than the environment itself. The world offered by the senses is only a possible world: it contains many more events that those to which the system can react to.

We can make examples even in the case of humans. We believe in a world in which we are possible causally connected to all kinds of visual and sensorial patterns. However, for a normal user it is impossible to recognize the difference between a Pepsi and a Coca. Many people are not able to read and, as a result, they are not able to recognize printed characters.

The following model is proposed. The environment is defined as the set of all possible events in the surrounding of a given agent (World 3). The sensory and motor capabilities define a second subset of the first one. Only those events that *can* produce an effect (directly or indirectly) in the agent by means of its body structure are part of the second subset (World 2). However this does not entail that the agent is causally related with all of them. This would the same as claiming that just because someone has a pair of ears s/he should be able to have a reaction for all possible kind of music and different language. This is not the case. Only a limited number of them (World 1) have been selected during the past phylogenetic and ontogenetic history of the agent and only those constitute the relevant environment for it (Fig. 2a).

A caveat is here needed. When we present these sets we are not referring to sets of objects. We refer to set of events. From the point of view of the user, s/he is causally related to events, not to objects. For instance s/he reacts to an execution of a symphony, s/he reacts to the showing of an image, s/he reacts to someone shouting. By means of the events the agent is usually able to relate with objects and other more or less static entities. However the first contact with the world is based on events, not on objects. It is arguable that objects are superset of events constructed on the basis of events [Manzotti and Tagliasco 2001; Manzotti 2003].

What do we mean here for an event to belong to the agent's world? The concept is straightforward. An event belongs to the agent's world if and only if, when it takes place, it produces a distinguishable effect in the agent. What do we mean for a distinguishable effect? A few examples are useful. If a flow of high energy x rays is projected against an unknown subject, it will produce with high probability several harmful effects on the cellular structure of his/her body. However, it would go unnoticed. If a pattern is showed to some uninterested subject it would produce effects (in the subject's retina a flow of chemical activity would surely be correlated with the visual property of the pattern), however it will have no relevance for his/her following activity. On the other hand if a new person is introduced, it will normally produce in a subject a lasting capability of reacting to his/her visual appearance. When a new word is heard, it produces a new capability in the subject. Namely, the capability of reacting to that sequence of phonemes. If 'qyxtzy' is heard, it does not produce any effect in English speaking listeners. If 'love' is heard, some kind of effect goes on.

What is the difference between the two? A first difference is that the couple (event and agent) must be able to produce a recognizable and repeatable effect. Is there any simple criterion to identify 'a distinguishable effect'? Yes there is. It is the fact that the system must be able to use that event as a goal. This is the crucial difference. When something happens, it must not only produce an effect in the agent, but the effect it produces must be usable as a goal for the system as a whole.

Let us go back to the previous examples. Accidental effects and unnoticed patterns cannot be used as goals by the system. Therefore they are not distinguishable notwithstanding the strength of the effects produced. On the contrary a new face or a new word can be used as goal. 'love' is a big goal for many, 'qytxtzy' is not. However it could become. If for example tomorrow a subject would read an interesting tale about the adventures of a curious character called 'qytxzy', s/he would remember it and s/he could use it as a new goal.

This last example is important because it brings us back to one of the original problems: where are the relevant events coming from? They come from the past of the history of the agent. They come from the environment itself. But how? We just said that the agent is enclosed inside a world of relevant events (world 1).

Well, the answer is that the boundary between the agent's world and the possible worlds is an open boundary. New events can drop from the world 2 in the world 1 by enlarging the world 1. Therefore a better metaphor would be one in which the boundary between world 1 and world 1 are not represented by a closed line but by a bubble like process of enlargement (Fig. 2).

A general schema is provided in Fig. 3. The world is perceived and acted upon through the motor and sensory bottlenecks. The agent perceives the world through the senses but it perceives only those events which are compatible with its perceptual categories. The matching between these perceptual categories and the current sensory input is fed into some behavior generator. In most cases of artificial agent this is what happens. However in an ideal system, the perceptual categories and the goals are the result of a process of epigenetic development. By the term 'epigenetic' we mean that it depends on the contingent experience the system had during its own lifespan. There are two situations in which the generation of goals is important. First, controlling the learning of correct behaviors with respect to the developed goals. Second, controlling the learning of new perceptual categories.

An ideal agent is capable of selecting whatever event from the environment (among those detectable by means of its sensorial capabilities) and then of using it as a goal for further development. Since the goal comes from the environment and since the capability of detecting it must be a result of the will to detect it as a goal, there is some kind of circular causality going on. It is not an undesired result. It is what we should expect since it is what happens during phylogenesis and ontogenesis.

A few words can be spent on this issue. The selection of new events in the environment (we always refer to the world 2 in the following paragraphs) is a process that can be performed in three different ways: totally unconstrained, partially constrained and completely constrained.

The last case is equivalent to most of artificial robotic implementations around. There is a fixed set of goals, hardwired at design time. The learning is aimed at achieving these goals. There is no space left for enlarging the repertoire of relevant events in the environment. The world 3 is empty and it is not enlarging. (In the world 3 there are only the newly acquired or self emerged events). The world 3 is missing.

The totally unconstrained case is of course unrealistic since it would lack any possibility of control. The outcome of the agent development would be completely unpredictable. The enlargement of world 3 is left to chance and to the resources available to the agent.

The partially constrained case is the most interesting for us. It leaves some space for the enlargement of world 3 but it will focus its enlargement on some pre-defined criteria. For instance human beings are going to devote their perceptual capabilities to recognition of human faces much more than shapes of rock or leaves. There must be some phylogenetic bias to focus their development on specific kind of events.

The agent will internally implement the process of self-emergence of new perceptual categories. How this can be done will be explained in the following paragraph. What is interesting is that the system will produce two outputs: a first perceptual output which represents the current sensorial input *after* it has been somehow matched with the epigenetic perceptual structure and the goal signal. These two signals have a completely different logical and practical meaning but are both the result of the epigenetic history of the agent.

4 An architecture capable of developing new goals

The architecture we present in this paragraph is a potential candidate to endorse the epigenetic development of new goals we just described. A detailed description of it and some experimental results can be found in Manzotti, Tagliasco 2005 [Manzotti and Tagliasco 2005].

There are probably other architectures with the same capacity. However what we present here must be seen as a valid candidate for the development of a completely adaptive agent (Fig. 3).

The main goal of the architecture is to let events to become goals for the systems. Every real system doesn't have necessarily to start from scratch. Some information can be embedded in the system in such a way as to permit the system to bootstrap itself and to take advantage more quickly of the environment. It's necessary to provide the system with some kind of meta-functions: a function that in a given environment will provide a specific function. In nature, an example of this meta-function is provided by imprinting. The bird has no specific knowledge of the shape of its mother, but has some meta-functions (looking for a moving object, in a fixed window frame of time). In this way the meta-function will be able to build a new function.

Since these functions must be ready before the development we called them 'phylogenetic'. They could have been as well been called 'hard-wired' functions. There are two kinds of phylogenetic functions: normal and meta.

Since the potential complexity of the environment is unlimited, it makes sense to have a mechanism in order to reduce the span of events to categorize and to explore. This mechanism corresponds to the phylogenetic functions. For instance, face recognition is triggered in human by some phylogenetic functions which is more interested to those events which are roughly similar to a face.

If an agent would have only phylogenetic functions it were an unsupervised learning or reinforcement learning system. It will have a set of goals fixed at design time: it will have a very limited capacity for development.

The epigenetic functions are equal to the phylogenetic ones. The only difference is that they are based on the categories developed during epigenesis. By using the categories developed during epigenesis, the system is partially unpredictable because, given a certain environment and certain phylogenetic functions, the agent will be able to pick up only certain events. However, in practice, given an unconstrained environment, the development will be unconstrained as well.

A system must be able to associate a value (proportional to how much that particular event is important on the basis of its past history) to what happens: this signal will be subsequently used as a reinforcement signal to implement the development of new goals.

Finally the system will produce, for selected classes of events, a relevant signal which is the maximum among all the phylogenetic functions and the epigenetic ones. It tells how much the current input is representative of the past history of the system. At the beginning, this signal will only be the result of the phylogenetic functions, but when new values come in from the epigenetic functions, the relevant signal will include the new goals and will allow the system to develop new goals.

Bibliography

R. C. Arkin, *Behavior-Based Robotics*, MIT Press, Cambridge (Mass) 1999.

V. Braitenberg, *Vehicles: Experiments in Synthetic Psychology*, MIT Press, Cambridge (Mass) 1984.

R. A. Brooks, C. Breazeal et al., 'The Cog Project: Building a Humanoid Robot', in Nehaniv (ed.), *Computation for Metaphors, Analogy, and Agents*, 1562, Springer-Verlag, Berlin 1999.

J. L. Elman, E. A. Bates et al., *Rethinking Innateness: A Connectionist Perspective on Development*, MIT Press, Cambridge (Mass) 2001.

K. Fukushima, M. Okada et al. 'Neocognitron with Dual C-Cell Layers', in *Neural Networks*, 7(1): 41-47, 1994.

S. J. Gould, *Ontogeny and Phylogeny*, Harvard University Press, Cambridge (Mass) 1977.

R. Manzotti, 'A process based architecture for an artificial conscious being', in J. Seibt, *Process theories*, Kluwer Academic Press, Process Theories: Crossdisciplinary studies in dynamic categories: 285-312, 2003.

R. Manzotti and V. Tagliasco, *Coscienza e Realtà. Una teoria della coscienza per costruttori e studiosi di menti e cervelli*, Il Mulino, Bologna 2001.

R. Manzotti and V. Tagliasco, 'From "behaviour-based" robots to "motivations-based" robots', in *Robotics and Autonomous Systems*, in press.

G. Metta, R. Manzotti et al., *Development: is it the right way towards humanoid robotics?*, ISA-6, IOS Press, Venezia 2000.

M. Ridley, *Nature via nurture*, Harper, Great Britain 2004.

R. S. Sutton and A. G. Barto, *Reinforcement Learning*, MIT Press, Cambridge (Mass) 1998.

T. Togawa and K. Otsuka, 'A model for Cortical Neural Network Structure', in *Biocybernetics and Biomedical Engineering*, 20(3): 5-20, 2000.

J. J. Weng, 'Cresceptron and SHOSLIF: Toward comprehensive visual learning', in S. K. Nayar and T. Poggio, *Early visual learning*, Oxford University Press, New York 1996.

Riccardo Manzotti and Vincenzo Tagliasco, DIST, Università di Genova

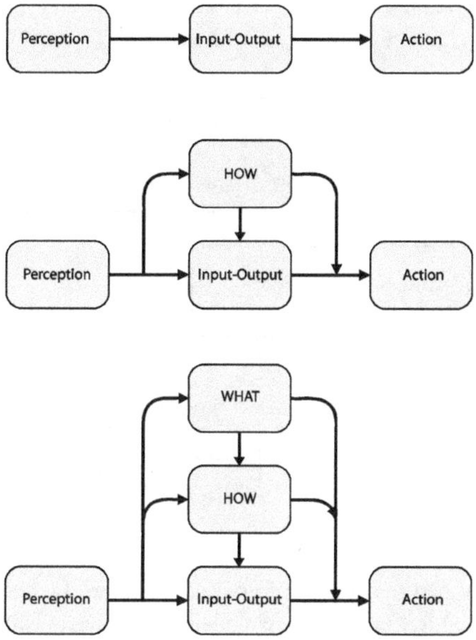

Fig. 1. Three possible architectures: top) both *what* and *how* the system does is defined *a priori*; middle) the system modifies *how* it behaves but not *what* is doing; bottom) the system modifies both *what* and *how* it does.

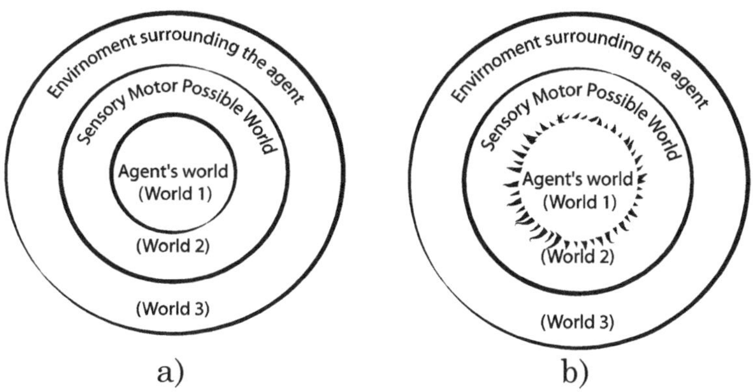

Fig. 2. a) The agent's world is a subset of events of all the possible patterns and combinations of sensory motor stimuli that are a product of the interactions with the environment. In turn the environment is an even larger set of events. b) The agent's world can expand.

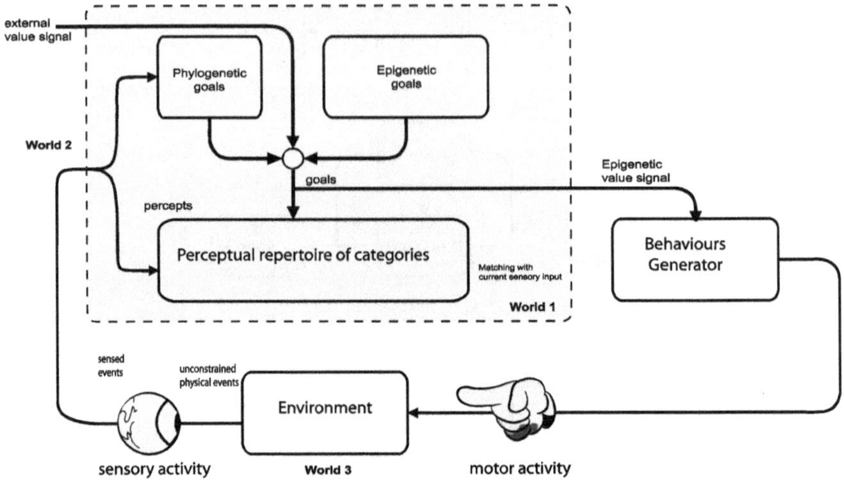

Fig. 3. A general architecture capable of goal generations. Goals are dependent on the perceptual repertoire which is in turn responsible for the events among which goals are selected. It is the kind of circularity which is responsible for the self emergence of goals.

Lella Mazzoli, Roberta Bartoletti
and Giovanni Boccia Artieri

The artificial in everyday life: social uses of communication technologies

Abstract:

The everyday artificial lies in the ambivalent relationship between a developing abstraction of complex society and the rediscovered importance of emotions as an inalienable characteristic of individual experience and the production of individual meaning.

The authors propose to analyze the everyday artificial by way of three keywords – immersion, contamination and contingency – providing various examples, from widely dispersed personal technological objects, such as mobile phones, to media walking, to affective computing, to intelligent machines, which are able to interact with their users and to learn from their relation with them.

Keywords:
social uses of technologies, emotions, immersion, contamination, contingency

1 New technologies, social systems and subjectivity: the artificial in everyday life

The advent of new technologies, which have become a distinctive and even unavoidable feature in today's complex social systems, offers us a unique opportunity to observe the changes underway and their underlying logic. This is particularly true for communication technologies, as our society is constantly redefining its identity, using communications as a starting point and fundamental process, both in the social and on a more individual level.

In order to gain a better understanding of the characteristics of this complex web of social and technological horizons, a good starting point would appear to be the decline of the great narratives [Lyotard, 1979], which have slowly become devoid of meaning, losing coherence and force as a source of norms. As a consequence, the contemporary social system and correlated individual experiences are characterised, at least on the surface, by an increasing degree of plurality and fragmentation.

With regard to single individuals, this has led to ever more impersonal relationships, which are often generalised and abstract. The ties of solidarity within the community typical of the pre-industrial era have been strained by the logic of modern society: strong ties have been substituted by weaker ones, or, better yet, by new types of relationships that are no longer as strong as those in the past, but which are nevertheless still effective in producing forms of collectivity [Mazzoli, 2001].

In keeping with these transformations are, on the one hand, 'reflux' strategies adopted by individuals in private relationships, which can be seen, for example, in a tendency to avoid becoming involved in socio-political activism – at least in its traditional forms.

A recent example can be seen in the innovative new forms of social and political participation now made possible by modern communication technologies, including the simplest forms, such as the short text messages (SMS) sent via mobile phones, which contributed to the mass turn-out of demonstrators in Spain in the days immediately following the terrorist attacks on commuter trains. Yet another example is the use of short text messaging to send messages in support of – or to campaign against – candidates on the eve of the European elections, as reported by the Italian newspapers.

At the same time, new ways of 're-discovering' the use of the senses as a means of expression and understanding have developed, in marked contrast to the hyper-rationalized forms that now appear to be characteristic of both social relationships and technology itself (or in any case of a version of technology which can be seen as rational and scientific in a traditional sense).

In other words, at the level of subjectivity we can observe a relative ambivalence with regard to the growing fragmentation of experience, as it becomes more and more abstract; this is in keeping with the transformation underway in social relationships, and the concomitant re-discovery of emotions as a component of individual human experience, understanding and making sense of individuals.

Modern technologies enter into the discussion at both of these levels and have resulted in what is in essence the artificial in everyday life, which emerges from the encounter between two aspects:

- an emotional aspect, which can be traced to the 'micro' realm, and to the dynamics of individual narration;
- a rational aspect, typical of the logic of 'macro' processes brought to life through the new media technologies.

Thus, it will be useful to observe the artificial in everyday life through three key words which are predominant in and characteristic of contemporary experience:

1. *immersion*: social relationships become a place where polysensory experiences are had, as demonstrated by the many neotribal forms of aggregation (at concerts, in discotheques, or among sports fans). The strength of these bonds is linked to the high degree of involvement or sensitivity, which may in turn be characterised by a certain weakness of the ties that results in the disbanding of the group when the bonding experience which created it is over;
2. *contamination* between personal abilities and the potential of the technologies, which is reflected by and made relevant today by a similar contamination or overlapping between the knowledge paradigms and cognitive models that characterise contemporary society. These ways of working can be seen in the worlds of fashion and design in general, as well as in that of new technologies;
3. *contingency* of identities and social situations; in other words, a sort of nomadism. This errant quest for experiences and knowledge [Maffesoli, 1997] has become widespread and a constant, thanks in part to the many possible means that can be used for instant communication, and as a creative response to a rejection of traditional forms of sedentary lifestyle invented in and promoted by the modern era.

The declension of these three keywords will be made clearer through the analysis which follows of several cases that exemplify the concept of the artificial in everyday life in relation to social dynamics, the emergence of new forms of experience – both collective and at the level of individual subjects – which are now contributing to redefine the very concept of the term 'user'.

2 Technologies for inclusion (but not only) in daily life: the mobile phone

The technologies used in daily life constitute a space which is influenced by complex and indeed ambivalent dynamics related to a wide range of topics. These include each subject's production of meaning and of individual identity, the emergence of new concepts of the collective and the reproduction of the social and of communication as a social process which is not linked to the specificity of individuals or to concrete forms of ties.

One particularly relevant example of this web of different aspects can be seen in the use of cell phones. It is not an overstatement to say that they are now as widespread as the television set, and that they offer an even greater potential for pervasiveness in our daily lives, in that they follow their user much like a shadow and have become in every sense a personal technology.

In an ambivalent way, while the cell phone on the one hand embodies subjective, emotional and specific instances, on the other hand it embodies social, instrumental and abstract meanings, that emerge explicitly if analysed using the three keywords proposed.

With regard to immersion, cell phones can be inserted in processes of emotional transformation by means of their integration into the family of personal objects. This is giving rise to a sort of humanization of this technology, as it moves through processes different than those for analogical transformation of the language of this medium, that has already been considered in pioneering studies on socio-technical paradigms and on interfaces [Mazzoli, 2001].

We refer in particular to ascribing of meaning to and personalisation of the mobile phone, first and foremost as an object and only then as a technology, thus transforming it from a generic object to an object of one's own, distinct from all other ones.

In other words, as a result of this process, mobile phones have become an object to which we become attached or fond of, and thus charged with symbolic worth that is exquisitely subjective and intimate.

Within this discussion of the emotional perspective, it is important to bear in mind one of the new frontiers in cell phone-mediated communication: multimedia messaging services (MMS). Multimedia messages carry digital images, and are thus visual messages and a sort of portable digital memory; as a result, they constitute succinct – but emotionally significant – forms of self-narration available for communication purposes, for use in the concrete construction of identity and relationships, particularly in relationships among members of the younger generations.

With regard to contamination or overlapping of personal skills and technological potential, cell phones instead show a high degree of homology with respect to the logic of contemporary social systems, in that they serve as a medium which helps us to cope with complexity and thus as a medium for inclusion. In other words, they prepare (that is: socialize) individual users to cope with the incessant need for selectivity in our highly complex social system and, at the same time, allow users to become accustomed to and accept the idea of inclusion in communication networks.

Indeed, mobile phones not only make possible but also reinforce the individual's experience of living in an 'always-on' world with its potential for virtually unlimited availability of communication in the face of variables such as time, space and even context. This potential availability, which is concomitant to and part of the very nature of possessing a mobile, results in expectations on the part of others that can transform this potential in a sort of obligation to be always connected and always available.

Finally, with regard to the idea of nomadism, mobiles can be seen as a privileged medium in defining what would otherwise be possible and the equivalency of different situations, different emotional contexts and different identities. Cell phones, in fact, have made both the weakening of roots in a given context and the growing, radical contingency of identity which characterises contemporary experience become part of our daily lives.

Indeed, they have led to an unprecedented weakening of the meaning of the specific contexts in which individual experiences take place – the 'here and now' (*hic et nunc*) – in other words, that which has been selected from among numerous possibilities with respect to what remains in the background or has been postponed but could be actual in the future.

The mobile phone exemplifies the weak difference between the present and the virtual, between 'here and now' and 'someplace else, otherwise' which can become actual instantly and without any apparent gap or discontinuity between it and the 'here and now'. As a result, the idea of 'here and now' becomes radically contingent upon it. In the same way, the concept of 'someplace else, otherwise' becomes more and more important.

Through the use of mobile phones, the abstract logic of contingency becomes a concrete practice of contingency itself.

3 Extreme consequences: the artificial in daily life and the interiorization of contingency

New technologies are entering our daily lives through a sensory or affective logic able, in virtue of experiences, to contribute to processes which attribute meaning. What's more, these processes socialize subjects in contemporary society with respect to the new reality which involves ever-more inclusive communication. Experiencing the artificial in everyday life means allowing oneself to be included in the social system's information network and experiencing its contingency.

1. A first process of inclusion produced by the artificial in our lives is characterized by a convergence between the ubiquity of technology and polysensory immersion in it. On the one hand, we are faced with an increase in all-pervasive and ubiquitous media technologies, and with an ever-greater integration of the artificial in our daily lives: examples range from the growing number of video monitors and touch-screens that are now widely used in public and work environments, to the ever-more widespread use of computers which are becoming less tied to fixed workstations thanks to the new wi-fi

connections. A further example is the widespread use of *walking media*, portable devices or tools which provide users with the possibility to establish advanced connections to the media system using mobiles, *personal digital assistants*, or even *wearable computers*, thanks to which the possibilities offered by computers and their connections can be integrated into clothing.

An example that is already possible today involves gloves able to read the smart tags on clothes at a distance. Each article of clothing sold can be labelled with its own *smart tag* containing a range of information on the fabric type, size, etc. Rfid (Radio frequency identification) systems can be connected to devices such as the glove described above: Motorola's Hearing Hand technology, for example, is able to identify the clothing which would suit us best, based on a series of data stored on personal tastes in fabrics, colors and costs: this is but one of the many prototypes of *wearable technology* which incorporate a computer in just a few centimetres of fabric, making it capable of conversing with other electronic devices via wireless connections. This integration of electronic devices in everyday objects represents one of the new frontiers in research, in that it will require changes in our way of conceiving – and using – clothing, which will not only reflect our tastes in fashion and comfort needs but will also have endless new functions.

The tendency of the artificial to become more widespread is closely tied to our perception of it as part of daily life, and thus as familiar and close to us, thanks to the proliferation of immersive interfaces able to direct the range of human senses within a technical-communication environment. The Virtual Reality technologies, which use a head-mounted display and data glove, are a state-of-the-art application that is still mostly confined to scientific circles. The real turning point will lie in the spreading of these devices to both business and leisure ambits, as in the possibility to explore real environments as 'virtual tourists', thanks to special glasses able to visually immerge us in a historical re-construction in a perspective of *augmented reality*. Such uses of this technology should pose no problems in everyday life but will be able to change our epistemology significantly.

2. A second characteristic is an emphasis on our sensory and emotional relationships with the artificial. This is the case with *affective*

computing, which applies technologies with which users relate and interact sensorially – in keeping with our recognised fascination for the artificial technologies described by Benjamin as a sort of inorganic sex appeal – and which are able to recognize and provoke emotions. These range from wearable technologies able to recognize the preferences of the user and to anticipate his/her needs, to veritable artificial beings able to interact with us. A good example is the Furby® toy, which is able to 'evolve' in its interaction with its child owner, emitting new sounds and behaving differently on the basis of this interaction. The point of tension in this evolution can be identified at the moment in which two Furby® toys are placed together in the same environment and interact (thanks to infrared light), to the point where they may exclude their owners: in this case, the artificial is communicating with itself, indifferent to the presence of a human being.

3. One last point regards the automation and intelligence of everyday objects, which are becoming ever-more able to interact self-sufficiently and introduce the aspect of control within the confines of what is familiar, in a perspective according to which this generalized control becomes part of everyday life [Foucault, 1975]. This is true, for example, in the case of domotics, as can be seen in projects foreseeing 'smart' homes able to adapt to the tastes, preferences and needs of their guests by learning to recognize habits and choose behaviours according to them (*e.g.*, to adjust the intensity of the lighting, immediately play favourite music over the home audio intercom system, recognise voice commands to open doors, etc.). It is clear that, seen in this perspective, we will become ever-more free as the artificial becomes better able to automate itself and monitor our home environment. This control exerted by the artificial involves dynamics that can also be applied to external spaces, such as supermarkets[1], and is invading spaces in our daily lives through cybernetic automation of household technologies. The multinational Electrolux has conducted ex-

1 The prospect of smart shopping carts able to provide information on the costs and characteristics of the products placed in them and to memorise purchases, so as to allow automatic payments and personalised promotional offers of products on sale that are proposed on a display screen, using a computerized one-to-one marketing strategy based on an analysis of the buying behaviour of the individual consumer.

perimentation in that sense and is already marketing a refrigerator known as a *screenfridge* which is able to scan its content and verify 'best-by'/expiry dates and decide whether to send an order to the local supermarket.

In summary, the extreme consequences of the artificial in our daily lives as analysed above lead us to three key issues that will offer significant challenges for future observation:

1. 'Virtualisation': the impact of the artificial on our daily lives is leading us to a new familiarity with the aspect of contingency as we learn, on the basis of a new virtual culture, to perceive and treat the world as always 'otherwise possible' [Boccia Artieri, 2004] and incorporate this state of perception in our sensory processes.

2. Monitoring/Control: with the advent of the artificial in our daily lives, we have moved beyond surveillance, and entered into a cybernetic world whose confines are becoming ever-more automated and which produces an internalization of the regulation and discipline of our bodies and communication.

3. The 'Post-user'?: we are moving from the idea of the individual subject as user of technologies towards a concept involving the participation of both mind and body in a technological environment in which we undergo experiences and co-evolve together with machines.

Bibliography

R. Bartoletti, *La narrazione delle cose. Analisi socio-comunicativa degli oggetti*, Angeli, Milano 2002.

R. Bartoletti, 'Oggetti dell'indifferenza. Oggetti mediali e logiche del sociale', in *Sociologia della comunicazione*, n. 35-36, Angeli, Milano 2004.

G. Boccia Artieri, *Lo sguardo virtuale. Itinerari socio-comunicativi nella deriva tecnologica*, Angeli, Milano 1998.

G. Boccia Artieri, *I media-mondo. Forme e linguaggi dell'esperienza contemporanea*, Meltemi, Roma 2004.

G. Boccia Artieri and L. Mazzoli (eds.), *Tracce nella rete. Le trame del moderno fra sistema sociale e organizzazione*, Angeli, Milano 2000.

M. Foucault, *Surveiller et punir. Naissance de la prison*, Gallimard, Paris 1975.

G. O. Longo, *Il simbionte. Prove di umanità futura*, Meltemi, Roma 2003.

J.-F. Lyotard, *La condition postmoderne: rapport sur le savoir*, Minuit, Paris 1979.

M. Maffesoli *Du nomadisme. Vagabondages initiatiques*, Librairie Générale Française, Paris 1997.
L. Mazzoli 'Le nuove frontiere del corpo e della mente', in *Pluriverso*, n. 4, Milano 1996.
L. Mazzoli, *L'impronta del sociale. La comunicazione fra teorie e tecnologie*, Angeli, Milano 2001.
M. Negrotti, *The Theory of the Artificial*, Intellect Books, Exeter 1999.
M. Negrotti, *Naturoids. On the Nature of the Artificial*, World Scientific Publishing Company, Singapore 2002.
G. Piazzi, *Il Principe di Casador*, QuattroVenti, Urbino 1999.

Graziella Mazzoli, Roberta Bartoletti and Giovanni Boccia Artieri, LaRiCA – Laboratorio di Ricerca Comunicazione Avanzata, Università di Urbino

Sabrina Moretti

Citizens, users, society: a sociological survey

Abstract:

The term 'user' is spreading from technological fields, such as computing, to various fields involving different types of interaction. The objective of our survey is to study whether this diffusion represents a simple linguistic substitution of other terms, or a deeper social and cultural transformation. In other words, we are interested in understanding whether the use of the word 'user' in common language is a manifestation of a particular social phenomenon that is currently developing, and is thus worthy of investigation. The context within which we are analysing the diffusion of this word is one that relates to the concepts of 'citizenship', on the one hand, and 'the market', on the other. It seems, in fact, that in the public field, at least, there is a progressive change of reference towards a market perspective, in which the citizen is being transformed into a consumer of certain services.

The results obtained in this survey show that the figure of the user *turns out to be connected to the emergence of the 'market' as a 'global' point of reference, but a market in which all goods are not material, but impalpable and indefinable within exact spatio-temporal boundaries, and in which it is often difficult to establish a boundary between he who supplies certain goods and he who makes use of them. We can note that the word 'user' is used above all in those social context characterised by ambiguous and plastic interactions. In a technological context, for example, the user of the Internet interacts with a machine, but he is not able to identify those who stand behind it, so to speak, because they flow quickly every time he carries out a new operation. Outside the technological context, the word 'user' seems to maintain this connotation that expresses the evanescence of those who participate in the interaction. In particular, in the opposition between public and private, the* user *seems to be he who uses a service supplied by an organization that is neither completely public nor completely private, but flows from one dimension to the other in an ambiguous manner. The sense of confusion and therefore of uneasiness implicit in the figure of the* user *seems to be confirmed by the fact that the majority of the interviewees did not attribute a positive connotation to this figure.*

Keywords:
user, citizen, global market, client, individualism, complexity, human-machine interaction.

1 User between citizen and client

The term 'user' is spreading from technological fields, such as computing, to various fields involving different types of interaction, in which the user is not only the user of machines, but also, for example, of services. The objective of our survey is to study whether this diffusion represents a simple linguistic substitution of other terms, or a deeper social and cultural transformation. In other words, we are interested in understanding whether the use of the word 'user' in common language is a manifestation of a particular social phenomenon that is currently developing, and is thus worthy of investigation.

The context within which we are analysing the diffusion of this word is one that relates to the concepts of 'citizenship', on the one hand, and 'the market', on the other. It seems, in fact, that in the public field, at least, there is a progressive change of reference towards a market perspective, in which the citizen is being transformed into a consumer of certain services. This gives rise to a change regarding rights and duties: the user is a rational being with more rights than duties, and he or she is able to evaluate the service received in relation to his or her needs and expectations. Along an axis of increasing abstraction, the concept of 'user' is tending to settle in an intermediate position between a set of contextually circumscribed and specific concepts, such as 'elector', 'taxpayer' and 'client', and a set of general and abstract concepts, such as 'person' and 'citizen'; but its significance is not yet clear with regard to the characteristics of the two sets of concepts. What *is* clear, however, is the pragmatic significance of the word 'user', indicating a relationship involving *someone* that uses *something*, wherein *something* can also indicate diverse institutions, structures, processes and social roles.

The research, from an empirical point of view, was carried out by means of a content-analysis of various documents taken from the most significant media sources, and through the distribution and subsequent analysis of a questionnaire. In this paper, we shall show some results obtained, by use of the questionnaire, from a sample of residents, chosen in a systematic manner from the electoral lists, from the town of Carpi (Modena, Italy).

The objective of this questionnaire is to reveal the use of the word 'user' in different situations, and in particular its relationship with other words, such as 'person', 'individual', 'subject', 'citizen', 'client', 'consumer' and 'taxpayer'. We suppose that all these words are positioned along an axis of abstraction upon which the word 'user' is placed somewhere near the middle.

The first part of this questionnaire includes 35 questions presenting specific situations, and subjects are asked which word – out of 'person', 'individual', 'subject', 'citizen', 'client', 'consumer' and 'taxpayer' – best represents someone who is in each particular situation.

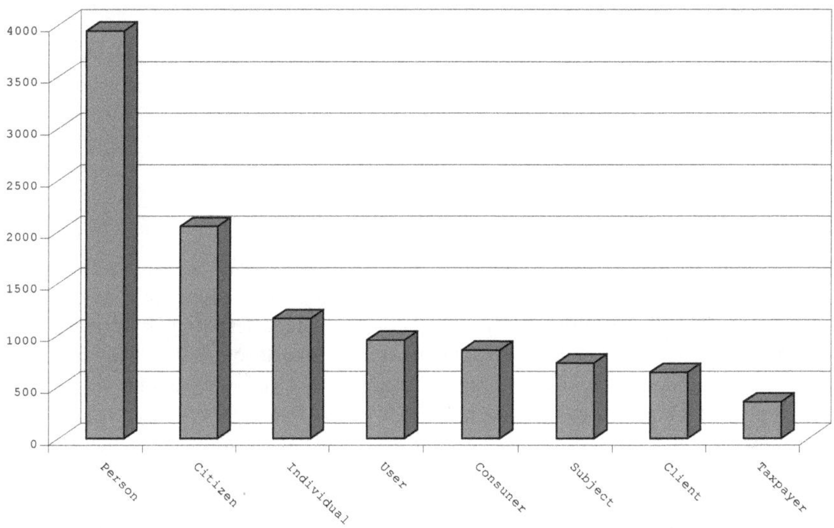

Fig. 1. Graph showing total frequencies of the 8 words.

If we analyse all of the answers (Fig. 1), we find that the more frequently used words are those characterized by a greater level of abstraction, such as 'person', 'citizen' and 'individual', while the words referring to a limited field, such as 'taxpayer' and 'client', are less frequent.

Next, we have applied a correspondence analysis[1] to the contingency table, which shows the frequencies of each term relating to every situation, in order to verify the existence of relationships among the categories of each variable, and whether there are some latent dimensions that are able to classify the data (Tab. 1).

1 The objective of the correspondence analysis is to describe the relationships between two nominal variables in a reduced-dimensional space, and at the same time to describe the relationships between the categories of each variable. This analysis proceeds by identifying latent dimensions through the analysis of the principal components, with the aim of creating new variables, obtained by means a linear combination of the original variables, so that a reduced number of these variables can 'represent' all the original data. The first component is the direction of the straight line (in a n-dimensional space, where n represents the number of variables to be reduced) along which the data can be distributed to maximize their variance; that is, the line along which they can be most widely scattered. We then project all of the data into the space perpendicular to the first component, and then repeat the process for each of the n variables, thus generating n perpendicular directions that are named either 'principal components' or 'latent variables'. The major part of the total variance is explained by means the first components. The correspondence analysis proceeds by simultaneously applying the principal component analysis to the matrix of row-profiles and column-profiles of two nominal variables.

Table 1. Contingency table showing frequencies of all words

Situations and activities	Modalities								
	person	citizen	consumer	taxpayer	individual	user	subject	client	Total
On the phone	87	11	33	2	21	119	19	11	303
On the motorway	55	18	42	10	18	90	13	60	306
Watching T.V. news	93	74	24	5	31	64	15	2	308
During a university lecture	121	15	6	7	65	29	51	6	300
Dealing with a public office	89	89	11	32	22	25	27	11	306
On holiday at the seaside	163	13	49	2	40	2	12	22	303
Listening to the radio	142	15	25	2	38	65	14	4	305
Going to mass	216	11	5	4	33	4	27	1	301
In a supermarket	29	9	168	7	5	3	2	79	302
Listening to a political debate	65	146	9	3	44	11	18	6	302
Taking the car to a mechanic	31	13	43	6	10	7	5	191	306
Dealing with the police	52	174	1	7	28	11	31	3	307
Navigating on the Internet	55	3	43	5	29	101	33	31	300
In front of the Gioconda	176	16	8	2	64	8	25	5	304
On the train	75	24	40	12	16	59	6	74	306
At school, during a parent-teacher meeting	204	30	1	8	25	4	29	2	303
Visiting a cemetery	228	16	4	0	36	10	12	1	307
Reading one's horoscope	131	7	23	1	76	10	54	2	304

continued overleaf

Situations and activities	Modalities								
	person	citizen	consumer	taxpayer	individual	user	subject	client	Total
During a visit to a show	135	23	19	2	73	19	28	6	305
During a doctor's visit	167	21	7	10	38	19	28	17	307
Renewing a passport	33	202	6	8	13	26	14	4	306
Reading a novel	183	9	41	2	40	8	21	2	306
Going to vote	28	255	5	8	4	2	1	0	303
During a city walk	188	58	3	2	41	3	11	1	307
During a tourist excursion	174	19	41	4	43	8	16	3	308
Play state lottery	67	7	97	47	25	15	17	28	303
During a country walk	221	17	3	2	45	1	15	2	306
During a civil suit	46	135	2	7	32	7	68	9	306
During a television program about the science	113	15	15	4	38	96	22	4	307
During a meeting with a councillor	65	180	4	17	14	8	9	10	307
At a concert	189	9	29	1	46	11	14	4	303
At a sports stadium	172	10	31	3	50	14	14	9	303
Appealing against a fine	46	161	4	23	23	11	37	3	308
Total	3939	2050	852	352	1161	951	727	636	10668

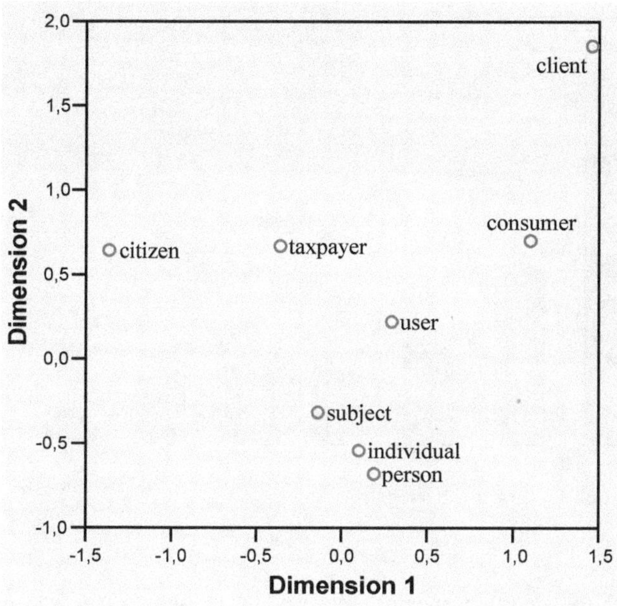

Fig. 2. Graphic obtained from the correspondence analysis.

If we analyse the graph obtained from the analysis (Fig. 2), we note that dimension 1 – which explains 37% of the variance – has at its extremities the words 'citizen' and 'client'. So, if we also analyse the distribution of the line points, we can assert that this dimension represents the opposition between public and private (the situations where the word 'user' is preferred are those related to a technological field and those in which there is an organization that distributes a service under a monopoly system, such as electricity, transport by train, etc., for which the distinction between public and private is not clearly defined). This opposition is also confirmed by the frequencies of the answers to the following question: *Each of us has daily relations with the State, its laws, its activities and interventions. In these situations how do you define yourself?*

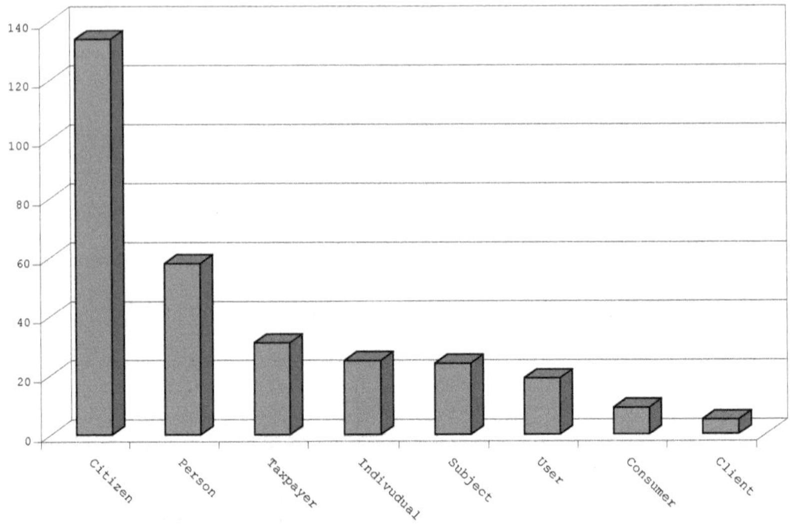

Fig. 3. Graph showing frequency of answers to the question: *Each of us has daily relations with the State, its laws, its activities and interventions. In these situations how do you define yourself?*

The results obtained (Fig. 3) demonstrate that if we order the eight proposed terms according to their frequency of occurrence, we get a scale that reintroduces the same order established by dimension 1 of our graph (with the exception of the word 'person', which shifts slightly).

Dimension 1, which demonstrates that the concept of 'user' is situated somewhere in the middle in the relationship between public and private, confirms the hypothesis that the diffusion of the word 'user' is linked to the gradual prevalence of the market as an ideal point of reference, where the concept of 'user' is in some cases replacing that of 'citizen'. The concept of 'user' seems to substitute the concept of 'client' in some fields that were originally public, while the concept of 'client' preserves its significance of purchaser of goods or services offered by a private provider.

The transformation of *citizen* to *user* and therefore to *client* of public services involves a variation with regard to the rights and duties of the 'recipient' of a service. We have surveyed the relationship between the concepts of 'citizen', 'user' and 'client', and the rights and the duties ascribed to each of them, by means of questions 49, 50 and 51, shown in Fig. 4.

In your opinion, how are the following figures situated as regards rights and duties?

	more rights than duties	same quantity of rights and of duties	more duties than rights
D49. Citizen			
D50. User			
D51. Client			

Fig. 4. Questions about rights and duties of all citizens, users and clients.

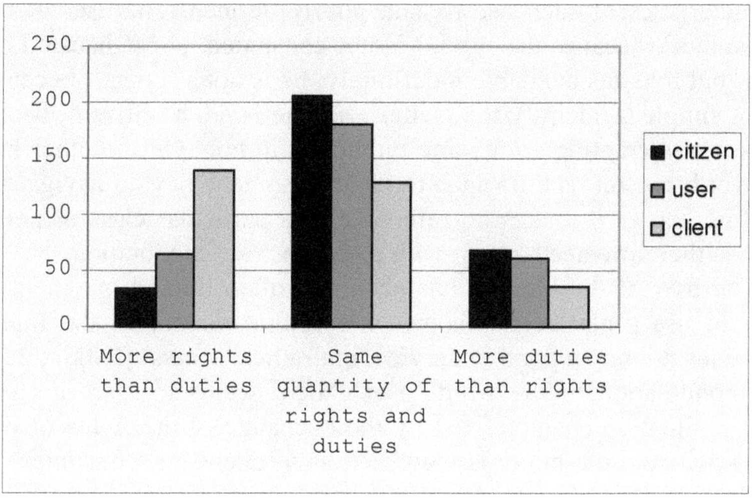

Fig. 5. Graph of frequencies of all answers to the questions in Fig. 4.

The results confirm the intermediate position of the concept of 'user' with regard to rights and duties associated with each figure, even if, from a statistical point of view, it seems nearer to the concept of citizen than to that of client.

Dimension 2 – which explains 30% of the variance – concerns, instead, the ability to identify an exact counterpart in each situation presented. Thus, the concepts of 'person', 'subject' and 'individual', which seem to have much the same conceptual connotation, are in marked contrast to the concept of 'client'. In those situations in which the word 'person' (or 'subject' or 'individual') is preferred, there is no direct or indirect interaction with some 'supplier', while, by contrast, we find certain situations in which the relationships involved, and the roles of the protagonists in these relationships, are well defined (e.g., when we take a car to a mechanic or go to the supermarket).

The location of the word 'user' in an intermediate position between 'client' and 'person' is easily verifiable in a technological field, from where the concept originates. A user of the Internet, for example, is someone who has acquired the right to use this particular tool, but the characteristics of such use remain poorly defined. The user is not a 'consumer', because the 'good' is not consumed, nor exhausted once used, but remains available indefinitely; he is not a 'client', because it is not simple to identify the 'seller'; and he is not a 'citizen', because he is not interacting with some public institution. On the other hand, the word 'person' is too vague to identify somebody who navigates on the Internet, as it makes no reference to a particular class of person, and we therefore need a term with a less abstract connotation.

The user of the Internet has acquired, often through payment, the right to use a particular resource. As Rifkin maintains, the Internet user does not buy a good or service, but rather 'access' [Rifkin, 2000]. The connotation of the word 'user', then, seems to mark a change from a situation characterized by the exchange of the rights of property – wherein the buyer is identified as a 'client' or 'consumer' – to another situation characterized by the exchange of the rights of 'use'.

The relationship between supplier and beneficiary is very vague, because the two actors cannot be identified by means of some procedure of standardization, and because, in some cases, it is impossible to trace a demarcation line between 'supplier' and 'beneficiary', as one can be both at the same time.

The concept of 'user', then, becomes the expression of a complex social phenomenon. This complexity is evident in the fields of computing and telecommunications, in which it is very difficult to design systems and user-interfaces that are able to respond to the needs of the *user* in an efficient and yet easy-to-use manner, given the complexity of the interactions developing during this process. To resolve this problem, the field of *user studies* has been born, the aim of which is to investigate some 'standard models' of users. The objective of these studies is to define some classifications of users in terms of their needs and expectations regarding the technological product being developed, in order to predict all possible requests that a user might make when using the product. The concept of 'user' indicates a generic social actor whose characteristics must be specified when an instance of the 'use' of a given resource is identified.

2 Individualization and complexity

The results obtained in this survey can help us to define the basis of a more extensive study for understanding the extent to which the concept of 'user' constitutes an enrichment, and thereby another manifestation of the process of differentiation of the social system, or whether, instead, it represents nothing more than a simple substitution of others concepts. In the first case, we need to investigate whether the diffusion of the concept of 'user' is connected to some transformations that are typical of modern society, and which are characterized by the emergence of some new types of social systems, such as the 'consumption society' or the 'information society'. The rapid growth of technology causes a cultural change whereby the 'user' represents a new figure, but we are still unable to identify which are those characteristics that clearly place him or her into the social system; and therefore it is not possible to define any associated behaviour and role-relationships from a sociological point of view.

By means of this survey, we can begin to define some questions relating to the diffusion of the word 'user', which can help us to formulate a hypothesis. The first conclusion that we may draw is that the concept of 'user' is involved in a process of individualization that characterizes modern society. According to Baumann, we are witnessing a slow but inexorable advance of private interest in a field originally managed by public power. The logic of the market is also invading all services that were part of the *res pubblica*, and a long process of privatization arises from this fact.

> The reproduction of the conditions of social-life is no longer attained by means of societal and collective tools, but, largely through privatization, has been removed from the domain of state politics and public decisions. 'Privatization' indicates in this case that the central powers are dispersing their responsibilities, and that the matters relating to social integration and to systemic reproduction are progressively being left to private enterprises [Baumann, 1999, p. 108].

Private interest, then, is prevailing over public interest, and so it is not important to realize processes and institutions that are able to create a collective of persons with some particular interest in common. In our society there are only very different individual interests which cannot be 'added together', so to speak, to give rise to a common cause. Anything that cannot be translated into the vocabulary of individual interests is no longer considered worthy of attention. Modernity marks a withdrawal from the *agorá*, the public space where the citizens discuss the common good. The new public space greater resembles a supermarket, in which everyone finds the 'goods' best suited to satisfying his or her own requirements.

The process of individualization of the modern society also entails a progressive crumbling of the concept of citizenship. Thus, in many situations, we do not feel like citizens, but simply individuals or clients. This aspect seems to emerge from our survey, in that we note, particularly among the younger participants, that the word 'citizen' is less often chosen than the words 'person' and 'individual', especially for those questions relating to the choice of an ideal figure.

The weakening of the figure of 'citizen' is also a consequence of the process of globalization, which entails a loss of power for governments, which remain local, in favour of a global power that is a long way from public institutions [Giddens, 1994]. The power is moving into the hands of the global market and financial capital, and it excludes all political agencies that remain localized. In this context, in which the links with the State seem to weaken, the concept of 'citizen' in the social interaction is less important than the concepts of 'individual' and 'client'.

The process of individualization, which for Baumann assumes a prevalently negative significance, can be seen in a more positive light when it attempts to pull down the 'iron cage' that imprisons the public services. The concept of a modern and rational state has led to excessive bureaucratization in all public services, so that their main activities are often limited to the execution of certain procedures that always follow the same well-defined rules, causing a homologation that does not consider the effective necessities of those who use these services.

The necessity to consider all individual needs has led to the attempt to 'humanize' all public services, and in particular the social services, designing into the public intervention

> the figure of *user*, borrowing from the market the concept of *client*; that is, he who goes to a place of distribution of goods or services to choose and consume those goods and services that he deems necessary for himself [Gui, 1995].

As in a market context – within which the main objective is not so much a meticulous development of procedures that regulate the supply of a service, but rather the satisfaction of a client – so also in all public services it is necessary above all to understand what the real necessities of each 'user' are, in order to be able to proceed to his full satisfaction. A user is not only a number but a person, with his own specific needs and expectations, which implies an adaptation of the supply procedures in the light of his individuality.

Our results show, however, that the concept of 'user' seems to give rise to a sense of confusion on the part of the beneficiary of a service, rather than the transformation from an anonymous citizen (marked by a number) to a humanized citizen (characterized by his own personality). It is true that this concept finds itself in an intermediate position between those of *citizen* and *client*, in both private and public contexts, but this location regards not so much the methods of each supply, but rather the identity of the provider of a service. In fact, in those situations in which someone interacts with a clearly defined actor, such as a public institution, the majority of our interviewees preferred to define him as a 'citizen', but they chose the option 'user' in all situations in which 'the other' is an organization that supplies some services under a monopoly (or quasi-monopoly) system, or an organization with public participation.

In all interactions with a public institution, there remains a sense of belonging to a community, connected to the concept of 'citizen'; in this context, everyone knows that a service is supplied in order to satisfy the exigencies of a collective. By contrast, the relationship between a client and a supplier of goods, or of a clearly private service, is regulated by the law of supply and demand under a system governed by competition, so that the client is he who can choose a supplier, and then choose to break off the relationship if it no longer satisfies his needs, or if the cost of the service is too high.

In those situations in which someone supplies a service that is not perceived as a 'common good', but is aimed, rather, to satisfy individual needs and exigencies, the recipient of this service loses the status of citizen. Then, if the service is not sold according to the market rules, which demand concurrency of various suppliers, it is not possible to call the recipient a 'client'. In this case, the concept of *user* is introduced, which as such becomes the expression of a situation of confusion and ambiguity. In fact, all the interviewees preferred to define as a *user* anybody who interacts with an organization whose public or private nature we are unable to identify precisely.

Another aspect that emerges from our survey concerns the concept of complexity connected to the diffusion of the term 'user'. This word finds its privileged context in the field of information technology. Above all, we note that the word 'user' is prevalent when we are not dealing with tangible goods, but with services. For example, when we use the Internet, we do not consume a good as such, but we 'access' a specific service; we buy the rights of use and not the rights of property. However, even in this case we can notice a difference from those who use some particular services, such as those offered by a lawyer, an accountant of other professionals, or a tourist agency. They do not sell something that is tangible, but offer their competence and knowledge to others, in order to resolve their problems and to satisfy their requests. In this case, we do not use the word 'user', but 'client' to name those who make use of such services. By contrast, those who use the Internet are not generally classified as 'clients', probably because the difference does not relate to the type of exchanged good, but rather the relationship with 'the other'. Who is the Internet? If it is anyone, it is a great many people who are invisible, both with regard to their identity and to the role they play in the interactions.

The Internet user is he who repeatedly uses a good characterized by a strong interactivity and complexity, in that it can be used in so many different ways. Information and computing technologies are used to read, to write, to find information, to calculate, to play, and so on. Thus the user can, at a given time, be a reader, a writer, a researcher, a player, etc. The word 'user', then, indicates a generic person that can assume any number of different roles that can be difficult to predict.

In a technological field, the complexity expressed by the word 'user' is related to the man-machine relationship. In this case, the term 'to use', which retains a rather generic semantic significance, is frequently adopted in reference to many issues relating to the man-machine interface: simplicity of use, flexibility of use, modality of use, etc. Another word introduced into the technical language is 'usability'. The engineering of *usability* is a discipline that has been present in computer science for some fifteen years, and it relies on vast numbers of research centres and university courses, as well as ever-continuing in-depth studies.

The mechanized exchange that allows us to obtain goods and services without the intervention of another human being poses, therefore, the sociological problem of the impact that it can have on social interaction [Margolin, 2004]. When we interact with a computer, the result that we obtain is a message that can be attributed neither to the seller of the machine, nor to the software programmer, nor to the feeder-in of the data It is the consequence of the work and the intervention of numerous social actors who remain anonymous, in that it is impossible to trace the identity of all involved. Esposito asks:

> With whom, then, has the user communicated, if nobody conscience can be intended plausibly as the sender of the information? With the computer? Is it necessary, then, to attribute to a machine intentions and motivations? [Esposito, 2003, p. 61].

From this point of view, the user of the Internet is he who establishes a new type of relationship with hidden and vanishing social actors who change each time he presses a key, and of whom it is impossible to grasp the identities, because they in continuous flux.

3 Conclusions

The results obtained from this explorative survey can help us to define some hypothesis that will subsequently be explored and perhaps verified.

Above all, the figure of the *user* turns out to be connected to the emergence of the 'market' as a 'global' point of reference, but a market in which all goods are not material, but impalpable and indefinable within exact spatio-temporal boundaries, and in which it is often difficult to establish a boundary between he who supplies certain goods and he who makes use of them.

Moreover, we can establish that the transfer of the word 'user' from a technological context to a much wider one is not concerned with its technical characteristics, but with the social complexity implicit in the type of interactions involved, which often amounts to a confusion of roles. The user of the Internet interacts with a machine, but he is not able to identify those who stand behind it, so to speak, because they flow quickly every time he carries out a new operation.

Outside the technological context, the word 'user' seems to maintain this connotation that expresses the evanescence of those who participate in the interaction. In particular, in the opposition between public and private, the *user* seems to be he who uses a service supplied by an organization that is neither completely public nor completely private, but flows from one dimension to the other in an ambiguous manner. The sense of confusion and therefore of uneasiness implicit in the figure of the *user* seems to be confirmed by the fact that the majority of the interviewees did not attribute a positive connotation to this figure.

In conclusion, the word 'user' seems to represent a 'manifold, complex world' that is in swift movement and therefore 'ambiguous', 'incoherent' and 'plastic' [Thrift, 1997, p. 178].

Bibliography

Z. Bauman, *La società dell'incertezza*, Il Mulino, Bologna 1999.

Z. Bauman, *La società individualizzata*, Il Mulino, Bologna 2002.

E. Esposito, 'Il linguaggio e i media della diffusione', in N. Addario (ed.), *La società come sistema autopoietico*, Carocci, Roma 2003.

A. Giddens, *Le conseguenze della modernità*, Il Mulino, Bologna 1994.

L. Gui, *L'utente che non c'è. Emarginazione grave, persone senza dimora e servizi sociali*, Angeli, Milano 1995.

V. Margolin, 'The Mechanization of Exchange', presented to the *5th International Conference of the Artificial: The User and the Artificial*, Ascona, Switzerland, 23-25 April 2004.

M. Negrotti, 'Appunti sul concetto di utente', in A. Porrello (ed.), Atti del Convegno *Modelli dell'utente nelle società avanzate*, Venezia 18-19 novembre 2002.

J. Rifkin, *L'era dell'accesso*, Mondadori, Milano 2000.

N. Thrift, 'The Rise of Soft Capitalism', in *Cultural Values*, 1997.

Sabrina Moretti, LCA – Laboratory for the Culture of the Artificial, University of Urbino

Teresa Numerico and Roberto Cordeschi

Meno's dilemma and web mining: the influence of search engines on the discovery process

Abstract:

WWW representation and organization of data is based on concept association, storytelling and interaction, while information retrieval techniques, which are normally used to search for and sort out information from the Web by search engines, imply instead a different model of cataloguing content: the database model. The organization of content that allows information retrieval means the creation of an interpretation and a transformation of the data into meaningful pieces of information, but at the same time causes the risk of misinterpreting them arbitrarily. It is therefore crucial to create innovative mining strategies for the Web, capable of taking into account its topological structure and its social and associative nature. This paper will discuss some of the problems raised by the massive use of search engines to gather information, and give a brief outline of some new perspectives on information access via the Web, using different searching strategies and paradigms.

Keywords:
search engines, Web information retrieval techniques, Web mining, semantic Web, machine-learning intelligent agents, link analysis, distributed IR, collaborating, filtering and categorization

1 Meno's dilemma of knowledge

The Web is not the only area in which it is necessary to solve the problem of correctly accessing and interpreting information in order to create new knowledge. This is, in fact, a very old philosophical issue that was discussed from the early stages of the history of thought.

Plato was well aware of the relevance of the topic in his epistemological research, as is clear from this passage of the *Meno* dialogue.

> MENO: But how will you look for something when you don't in the least know what it is? How on earth are you going to set up something you don't know as the object of your search? To put it another way, even if you come right up against it, how will you know that what you have found is the thing you didn't know?
>
> SOCRATES: I know what you mean. Do you realize that what you are bringing up is the trick argument that a man cannot try to discover either what he knows or what he does not know? He would not seek what he knows, for since he knows it there is no need of the inquiry, nor what he does not know, for in that case he does not even know what he is to look for. [Plato *Meno*, XIV 80d-e/81a.]

The dilemma is set up very clearly here: it is not possible to discover what we already know, because in that case there is nothing new, and we cannot discover what we do not know, because we have no clue whatsoever about how to understand that it is what we are looking for. The solution proposed by Plato is that even if we don't know things, there is a sort of pre-comprehension of what we are looking for, due to our previous *hyperuranic* experience of ideas in themselves. Though we have forgotten it, we still keep a vague memory of ideas that allows us to recall the previous knowledge when dealing with new experiences and discoveries. This is a very rough presentation of the Platonic world of ideas, and here we not wish to delve into the issue any further. We can approximately interpret the Plato/Socrates solution for our purposes as the assumption that when we are in the process of discovery, we normally do not start from scratch, but we use some prejudgemental model of the data to evaluate the results, and our knowledge-creating procedure consists of fitting our model into the data in order to make sense of our hypotheses.

2 The Memex project by Vannevar Bush

Vannevar Bush was one of the first scientists who considered the issue of information organization and retrieval as a key question. In 1945 he launched the proposal of the creation of a Memex, a device based on a totally innovative cataloguing method to retrieve data. The project relied on his intuition that our memory does not use the standard indexing practices used in archives and databases. His suggestion took place at the very beginning of the information and communication revolution of digital-electronic technologies. He was, at that time, one of the most influential scientists in the U.S., because President Roosevelt appointed him director of the Office of Scientific Research and Development (OSRD), a special agency reporting directly to the White House. Though he was the inventor of the Differential Analyzer, a very powerful analog calculating machine built during the 1930s, he was not able to foresee the importance of digitization of information. However, in 1945 he wrote:

> Our ineptitude in getting at the record is largely caused by the artificiality of systems of indexing. When data of any sort are placed in storage, they are filed alphabetically or numerically, and information is found (when it is) by tracing it down from subclass to subclass [...].
>
> The human mind does not work that way. It operates by association. With one item in its grasp, it snaps instantly to the next that is suggested by the association of thoughts, in accordance with some intricate web of trails carried by the cells of the brain [Bush 1945, Sect. VI].

Bush mentions here one of the most important and peculiar features of the storage of hypertextual information: the idea that the connections among documents are not related to a precise structure or index, but are generated by means of a case-by-case strategy which is not established in advance, independently of the content of the documents. He considered association and similarity as useful and efficient methods to retrieve remarkable information. The Memex, although it was never built, had the clear goal of escaping from the boundaries imposed by the traditional indexing procedures. The WWW is to some extent the achievement of Vannevar Bush's dream.

3 Web topology is not democratic

The Notre Dame University research group, under the guidance of Albert-László Barabási, set up an experiment to map the Web. Using methods commonly adopted in statistical mechanics, they analysed only a small fragment of the Web and then hypothesized the behaviour of the network as a whole, founding the conclusion on the assumption that it would have a similar topology in terms of connections of nodes. Through this experiment they obtained an astonishing result: on average there were 19 degrees of separation between two randomly taken Web pages [Barabási 2002, pp. 32-34]. As Barabási clearly stated:

> The most intriguing result of our Web-mapping project was the *complete* absence of democracy, fairness, and egalitarian values on the Web. We learned that the topology of the Web prevents us from seeing anything but a mere handful of the billion documents out there [*ibid.*, p. 56].

The fact that all information is theoretically available on the Web is not a guarantee that it is easily accessible. According to the latest results of the theory of the networks, it seems that social networks (and the WWW is not an exception) are dominated by a few highly connected nodes, the so-called *hubs*, which can be considered as the strongest argument against the utopian vision of cyberspace as an egalitarian arena.

4 The success of Google, and its faults

Considering all these characteristics, users could experience many difficulties when trying to orientate themselves towards finding useful or relevant information. The organization and the finding of pages can frequently be frustrating, due to the overwhelming and chaotic load of available data. That is why search engines play the role of mediators for accessing information, providing the 'guided tours' to visitors, by providing easy access to certain routes, whilst effectively concealing others. Users are very likely to obtain the information they seek via a search engine, and if a page is not listed in the results of a query, its location could remain unknown forever. This gives an enormous power and a huge responsibility to these means, because they offer the *unique choice* of results, outside of which there is only chaos and incomprehensible noise. The questions that arise spontaneously from these premises are: how fair is this Web 'guided tour', considering that we rely completely on it? Is a search engine presentation of the Web trustworthy? Could it be biased for commercial reasons or failures in representing the integrity and the complexity of the virtual world?

We concentrate on Google, which is the most successful search engine so far, trying to understand the reasons for its power in retrieving data on the Web. It is well-known that its efficiency depends, beyond the huge number of indexed pages[1], mainly on its search algorithm, PageRank, which is based on the exploitation of interconnections of Web pages, as a way to attribute authority to pages. This algorithm mixes a standard inverted index of all pages, treated as vectors of strings, with a system that allows it to attribute authority to a page according to the links that it receives from other pages. A link from page 'A' to page 'B' is interpreted as a 'mark' given to the 'B' page from the 'A' page. The value of the link depends also on the authority of 'A'. So the success of Google is related to its capability of taking into account the real topology of the Web when creating the ranking for the pages [see Brin and Page 1998]. Pages linked by a lot of authoritative pages have more credentials to be considered relevant for the users. However, even if the system is perfect when the user is looking for hubs and 'mainstream' pages, Google results are less trustworthy when we look for 'minority pages'. When we seek information that is not very well known, or not very popular, or even only written in a language that is not widely spoken – such as Italian, for example – it is much more difficult to obtain it [see Cho and Roy 2004]. Though the protection of minorities is not at the center of the

1 According to the latest estimate of the Google technical team, there are almost 8 billion pages indexed in the repository (10 November 2004, <http://www.google.com/googleblog/2004/11/googles-index-nearly-doubles.html>).

users' attention, it is a very relevant issue in the perspective of the creation of an ecology of the Web as a collective and distributed information tool. Thus, the ranking mechanism that is the main reason for the 'fitness' of Google's performance produces, at the same time, discrimination for all sorts of minorities present on the Web (language minorities, scientific minorities, young communities, etc.).

The visibility of minorities is not the only problem that we encounter when we use Google as our privileged door to the Web; there is also the Google bombing phenomenon. As an example, try to type the string 'miserable failure' into the query bar, and you obtain as the first result President Bush's webpage on the White House website (<http://www.whitehouse.gov/president/gwbbio.html>). This is the outcome of the exploitation of PageRank algorithm characteristics by the hackers. To achieve this result they do not need to enter in the target page. They only need to create a lot of pages, widely cross-linked by each other, using that term and connect them to the Bush homepage. Notice that this activity is perfectly legal. And it is not an isolated episode; all commercial companies can make the most of it to increase the ranks of their homepages relatively to the pertinent keywords [see Gori and Numerico 2003].

5 Database information retrieval versus association retrieval strategies

Retrieval of information has always been related to structured data. The key process of database creation is the interpretation of the data, in order for them to be organized in a precise way. The basic model for database creation is the transformation of rough data into meaningful content, and the procedure to achieve this goal requires the construction of a suitable framework for the data. The first priority of databases is to respond quickly, efficiently and coherently to queries by the users. The organization of content is subservient to the sorting out of information. Considered from this point of view, the creation of a database is the transformation of the data into a meaningful whole. When the Web became the biggest ever repository of data in the entire world, the only well-known methods to retrieve data came from the experience of large databases, as is clearly pointed out by some of the major experts in the area:

> Many of the search engines use well-known information retrieval (IR) algorithms and techniques. However, IR algorithms were developed for relatively small and coherent collections such as newspaper articles or book catalogues in a (physical) library. The Web, on the other hand, is massive, much less coherent, changes more rapidly and is spread over geographically distributed computers. This requires new techniques or extensions to the old ones, to deal with gathering information, making index structures scalable and efficiently updateable, and improving the ability of search engines to discriminate [Arasu *et al.*, 2001, pp. 2-3].

We can conclude that, at the moment, Web search techniques are neither always adequate nor always capable of taking into account most of the peculiarities of the complex data organization of the network, and that they risk misinterpreting information and concealing some relevant data so that they will remain undiscovered for the unaware user[2]. Although Web pages have a tag structure that allows a sort of database organization for information, in fact, the Web's complex topology and structure cannot be correctly and exhaustively described as a traditional database[3]. The associative and narrative character of the World Wide Web hypertext, as already anticipated by Vannevar Bush's vision, cannot be fitted into a database without loss of information, being only an abstract snapshot of the content that cannot preserve its integrity.

Moreover, we cannot underestimate the fact that research is a unique term used to identify a variety of activities, such as search through keywords, different methods to analyze structured and nonstructured data, flexible ranking mechanisms, evaluation of peer results, content organization, automatic rule-based machine-learning classification, relational taxonomies, taxonomy generation, social knowledge management, adaptive ranking based on social choices,

2 The discussion of the Invisible Web is outside the scope of this paper. For further information, see Sherman and Price 2001.

3 For more information about the Web topology according to power law degree distribution, see Faloutsos et al 1999.

etc. [Bawa *et al.* 2003]. All these different processes not only require diverse strategies to be properly performed, but are aimed at generating complementary results. For these reasons, no matter how efficient a search engine is, it is highly recommendable to project a wide range of means to gather data from the Web. In the following sections we present briefly some of the most promising directions of research in the field of Web searching.

6 Semantic Web (SW)

> The main idea behind SW is that we try to extend the current Web with additional information to make it possible that computers do understand the content of web pages. Of course, we are not going to remove or replace the current Web. The SW is an extension, an additional layer on top of the already existing Web [Van Harmelen 2003, p. 2].

The basic idea of SW is to add a layer to the currently existing Web, which provides users with extra information about the content and the meaning of Web documents, so that everybody can more easily access what they are looking for. Though this project seems to be very promising for restricted parts of the Web, such as good categories for e-commerce or library catalogues, it seems to be at least problematic, from an epistemological point of view, to create a standard vocabulary for metadata that is suitable to be applied to every area of knowledge. The creation of ontologies is a very tough task, and nobody knows if the effort will ever end with a consistent shared result. Moreover, it is not clear at the moment if machines should be able to attribute semantic tags directly to data, or if we need a group of people with the mentality of librarians, who will add metadata information to all the resources in a normalized way. Both solutions, although involving different difficulties, seem to share the characteristic of being a huge amount of work that could take years to carry out, and which risks requiring a never-ending checking process. If we look at SW according to Tim Berners-Lee's vision, it will 'add logic to the Web' [Berners-Lee, 1998]. So, even if we suppose that it is possible to overcome all the various practical obstacles, and to achieve the desired result, what we will end up with is something that is not similar to the Web, as conceived by Berners-Lee, which we have learnt to appreciate, but a completely new system. It will become a semantic-oriented database, and it will lose the associative and innovative links opportunities anticipated by Vannevar Bush, being restricted to the logical structure imposed by the metadata layer that is not capable of taking into account the serendipitous connections between different topics.

7 Distributed Web mining

This research area is many-sided, so that we can identify various streams that belong to it in one sense or another, but which all share the view that a unique, centralized data repository management is not the best way of investigating the variety of content on the Web. They also mainly rely on the idea that machine-learning bottom-down strategies can help both the indexing process and the personalization of results according to users' preferences. We can distinguish at least three different approaches:

a) Deep agents and focused crawling strategies
b) Shallow agents[4] and personalization issues
c) Peer-to-peer (P2P) clustering creation

a) Deep agents
These are intelligent agents that surf the Web, not on a random basis as Google crawlers do, but by following a topic-orientation strategy in choosing the links to follow. We consider these intelligent agents 'deep' because they work at the crawler level and create their own index without relying on a unique repository. They could use PageRank or another link analysis algorithm, but their most important feature is the integration of the crawling activity with the evaluation and clusterization process of Web pages [see e.g. Diligenti *et al.* 2000 and Chakrabarti *et al.* 2001][5].

4 We rely on Gori and Numerico, 2003, for a discussion on the adopted terminology.
5 For a discussion of all practical issues of implementing such a strategy, see the focuseek project (<http://www.focuseek.com>).

b) Shallow agents
These agents operate on a collection of pages already provided by a classic search engine. The agents are user-oriented and organize information on the basis of implicit or explicit knowledge that they gather about the user's preferences. Vivisimo (<htpp://vivisimo.com>) meta-search engine, for example, organizes the results of each query adopting a clusterization method that avoids the scanning of many pages for the desired response. The integration of deep and shallow agents will allow the realization of a really personalized and distributed search strategy. The topic-oriented agents will create the indexed collection of results that will be presented to the users according to their implicit or explicit range of preferences. These new techniques, once completely developed, will eventually be able to face the monopolization issues raised by a centralized unique repository of documents [see Gori and Numerico 2003][6].

c) Peer-to-peer clustering creation
Distributed Information Retrieval (DIR) studies problems of searching distributed and heterogeneous repositories. In order to obtain results using these techniques, there are various issues that need to be tackled: the identification of a common language for the resource description, the creation of repository selection schemes, and the development of methods for merging different ranked lists returned from searches in non-homogeneous archives. P2P networks are distributed and dynamic; they allow the participation of many members and the use of low-cost systems, built on the principle of utilizing unused computing resources.

SETS (Search Enhanced by Topic Segmentation) is a very promising project in this area, which puts together the machine-learning traditions of topic segmentation studies and social network results, integrating them with P2P search methods [Bawa *et al.* 2003]. SETS is in fact a topic-segmentation-based network that provides efficient search facilities on P2P networks. The key idea is to arrange sites in a network such that a search query probes only a small subset of sites

6 For further arguments in favour of a distributed Web search architecture, see Khoussainov *et al.*, 2001.

where most of the matching documents reside. In the network, there are two kinds of links: short-distance links connecting sites within the same content segment; and long-distance links connecting pairs of sites from different segments, so that queries are routed only in the relevant areas of the network, without flooding the whole network with irrelevant traffic.

8 Collaborative filtering and categorization

The fact, already stressed, that the Web can be considered a social network, can be exploited also on the side of potential explicit or implicit social collaboration in the organization and categorization of the network information. If we can create devices that allow us to share and organize collectively the knowledge of each individual member of the community, and if we succeed in creating methods to propagate trust and to control distrust [see Guha *et al.* 2004], then we can benefit from the expert evaluation of links, and from the creation of a collectively built, trustworthy, free knowledge base[7] [see, e.g., Resnick *et al.* 1994].

In social data-mining systems, which represent an enhanced version of collaborative categorization, it is not even required that the users be engaged in an explicit contribution to the community. The system seeks to exploit users' behaviours to detect the information implicit in the records of their existing activities [see Amento *et al.* 2003].

A bright and successful example of implementation of these techniques is the Amazon knowledge-management system, that uses the buying habits of previous clients to provide prospective customers with suggestions that are frequently very effective, relying on preferences of other, similar, customers. The overall system also benefits from the creation of the sense of belonging to a virtual community of shared interests.

7 See, for example, the Internet Public Library (<http://www.ipl.org/>), a system that allows every member of the community to contribute to the categorization and the organization of websites and that relies on the reciprocal trust of members.

The hypertextual, distributed and associative nature of the Web implies the necessity of innovative research strategies in order to overcome the weaknesses and shortcomings of the current information-retrieval techniques. Distributed and social systems of data mining offer promising approaches to Web search techniques and methods. Whatever the future of Web searching will be, however, it is important that the user always be aware of the difficulties and of the delicate intermediation role played by search engines in the process of accessing information and in the consequent outcomes in terms of knowledge creation.

Acknowledgements

We would like to thank Jonathan Bowen for his invaluable help, and for the many useful suggestions that improved the paper. Teresa Numerico wishes to thank the Leverhulme Trust for its support during the writing of this paper.

Bibliography

B. Amento, L. Terveen, W. Hill, D. Hix and R. Schulman, 'Experiments in social data mining: the TopicShop System', in *ACM Transactions on Computer-Human Interaction*, 10, 2003, 1, pp. 54-85.

A. Arasu, J. Cho, H. Garcia-Molina, A. Paepcke and S. Raghavan 'Searching the Web', in *ACM Transactions on Internet Technology*, 2001, 1, pp. 2-43.

A. L. Barabási, *Linked*, Perseus Publishing, Cambridge MA 2002.

M. Bawa, G. Manku and P. Raghavan, 'SETS: Search Enhanced by Topic Segmentation', in *Proc. of the 26th Intl. ACM Conf. on Research and Development in Information Retrieval (SIGIR)*, 2003. <http://citeseer.ist.psu.edu/bawa03sets.html>.

T. Berners-Lee, *Semantic Web road map*, 1998. <http://www.w3.org/DesignIssues/Semantic.html>.

S. Brin and L. Page, 'The anatomy of a large-scale hypertextual web search engine', in *Computer Networks and ISDN Systems*, 1998, 30, pp.107-117. <http://citeseer.ist.psu.edu/brin98anatomy.html>.

V. Bush, 'As we may think', *The Atlantic Monthly*, July 176, 1945, pp. 101-108, <http://www.ps.uni-sb.de/~duchier/pub/vbush/vbush-all.shtml>.

S. Chakrabarti, M. van den Berg and B. Dom, 'Focused crawling: A new approach to topic-specific web resource discovery', in *Proceedings of the 10th International World Wide Web Conference*, Vol. 3, ACM Press, Hong Kong 2001.

J. Cho and S. Roy, 'Impact of search engines on page popularity', in *Proceedings of the WWW2004*, ACM, New York 2004, pp. 20-29.

M. Diligenti, F. Coetzee, S. Lawrence, C. Giles, and M. Gori, 'Focused crawling using context graphs', in *Proceedings of the 26th International Conference on Very Large Databases, VLDB* Vol. 3, El Cairo 2000.

M. Diligenti, M. Gori, and M. Maggini, 'Web page scoring systems for horizontal and vertical search', in *Proceedings of the 12th International World Wide Web Conference*, Vol. 3, ACM Press, 2002.

M. Faloutsos, P. Faloutsos and C. Faloutsos, 'On Power-Law relationships of the Internet Topology', in *Proc. of ACM SIGCOMM*, Aug. 1999, pp. 251-262. <http://citeseer.ist.psu.edu/michalis99powerlaw.html>.

M. Gori and T. Numerico, 'Social network and web minorities', in *Cognitive Systems Research*, 2003, 4, pp. 355-364.

R. Guha, R. Kumar, P. Raghavan, and A. Tomkins, 'Propagation of trust and distrust', in *Proceedings of the WWW2004*, ACM, New York 2004, pp. 403-412.

S. Lawrence and C. L. Giles 'Searching for the World Wide Web', *Science*, 3 Apr. 1998, 280, pp. 98-100.

R. Khoussainov, T. O'Meara and A. Patel, 'Independent proprietorship and competition in distributed Web search architectures', in *Proceedings of the Seventh IEEE International Conference on Engineering of Complex Computer Systems* (ICECCS), 2001. <http://citeseer.ist.psu.edu/660982.html>.

Plato (*c.* 380 B.C.), *Meno*, translated by Benjamin Jowett, <http://classics.mit.edu/Plato/meno.html>.

P. Resnick, N. Iacovou, M. Suchak, P. Bergstrom and J. Riedl, 'GroupLens: an open architecture for collaborative filtering of networks', in *Proceedings of CSCW'94* (Chapel Hill, NC), ACM, New York, November 1994, pp. 175-186.

C. Sherman and G. Price, *The Invisible Web: Uncovering Information Sources Search Engines Can't See*, Information Today, Medford 2001.

Van Harmelen, 'The future of search engine "search and you will find". An interview with Franck Van Harmelen', in *Network*, 2, 2003, pp. 1-6. <http://lgxserve.ciseca.uniba.it/lei/ai/networks/03-2/vanharmelen-en.pdf> (both hyphens included in URL).

Teresa Numerico and Roberto Cordeschi, Dipartimento di Scienze della Comunicazione, Università di Salerno

Giacomo Romano

The problem of the individuation of artifacts

Abstract:

Many studies in the cognitive sciences aim at an understanding of how human categorize and conceptualize items in the world. Artifacts are interesting subjects of investigations for these disciplines, as often, in the human conceptual system, are opposed to natural items. However, cognitive research seems to have neglected the complexity of the task that human engage in categorization and conceptualization of artifacts. In my paper I point to a difference that has been little taken into consideration. This is the difference between the process that I call the individuation of an artifact, by means of which one recognizes that a certain item is an artifact. And the process that I label as the identification of an artifact, by means of which one recognizes that a certain item is an artifact of a certain kind.

My intention in this paper is to point at a theoretical problem concerning the possibility of recognizing an artifact without recognizing what kind of artifact it is. It seems that this problem has not been taken much into consideration by the cognitive sciences, which have been interested with other issues involving the human conceptual system.

Keywords:
artifacts, artifact individuation, artifact identification, artifact conceptualization, design stance, means-end reasoning, teleo-instrumental attitude

1 The Ambiguity of 'Artifact'

A short summary of the literature in cognitive psychology and neuropsychology [as for cognitive psychology cf. e. g. Bloom 1996; Landau, Smith, and Jones 1998; Ahn 1998; Gelman, and Bloom 2000; Matan, and Carey 2001; as for cognitive neuropsychology cf. e.g. Biederman 1987, Martin, Ungerleider, and Caramazza 2001; Mahon and Caramazza *in press*], involving the study of artifacts, shows how much they do need a boost of conceptual analysis. In fact, each article about artifacts does not distinguish between artifacts, which are instruments, tools, or other items. Often the artifacts taken into consideration are objects to which a function can be ascribed without particular problems. Also when an 'artifact' is used as synonym of 'human made thing' (especially in neuropsychology), it refers always to an object with a specific, defined function. This fact is sometimes driven by the nature of the experiments. When the subject matter of an experiment is, for example, a categorization task, the function[1] of an artifact must be as recognizable as possible.

According to me, there is an extremely compromising assumption that already biases the way in which we conceive of artifacts. If one takes for granted that the function of a certain artifact is recognizable, overlooks a basic step in the procedure that makes us recognize the general class of artifacts. It seems that in categorizing an artifact one takes two steps, which are logically distinct. These consist of processes that I name *individuation* and *identification*, and which, I claim, are distinct. It likely appears that one first understands that an item is an artifact (*individuation*) and then understands that it is a certain kind of artifact (*identification*). Apparently, before understanding that a corkscrew is an object that works to screw corks, one understands that a corkscrew is something created by a human-like manufacture.

Almost all the cognitive studies about artifacts have been focussed on the procedure of their categorization. They have pointed at the results of this cognitive activity: they have looked only at the cognitive taxonomies derived from it. The necessary requirements for the categorization of artifacts seem to presuppose the process of identification. In fact, according to the data derived from the research of cognitive psychologists, the procedure of categorization is based on the recognition of three salient features. These are, in order of importance: 1) the function for which an artifact is designated (by its creator, or by its maybe user); 2) the category to which an artifact was designated to be-

1 In contexts like this I think that the notion of 'function' can be explained as the proper use for which a certain artifact is designed. Also in this case, however, things are not clear.

long by its creator; 3) the material and the physical structure which give to the artifact its shape (cf. the references above mentioned). These three features can overlap, of course, in the artifacts, which fill our daily life.

2 Artifacts Individuation and Identification

The procedure of artifact categorization requires the capacity of recognizing these three mentioned basic features. Such ability has been taken as basic and constitutive, but only during the last few years a few psychologists have taken the challenge of investigating the cognitive processes underlying it. The three recognition processes composing the procedure of categorization are probably constitutive of the psychological attitude that has been described by Daniel Dennett [1987] as the *Design Stance*. This is:

> … an abstract explanatory schema that captures the relationship between features of an entity (e.g. its material, shape and activities) in terms of a coherent organizing notion – the purpose for which its designer created it [Defeyter and German 2003, p. 134].

The *design stance* is *an abstract explanatory schema* defined in *folk psychological* terms. However, the psychologists and philosophers who refer to it, as well as Dennett himself [cf. also Dennett 1990], have not clarified what exactly the *design stance* consists of. The task of explaining it may plausibly represent a wide and interesting field of research for cognitive interdisciplinary studies, and it can stimulate interesting insights in the understanding of artifacts. But there seems to be a logically prior question. In fact, even in appealing to the *design stance* as a heuristic and descriptive instrument, an individual already presupposes the capacity of recognizing an artifact and of ascribing a function to it. And the process of recognition is yet completely unexplained.

The (analytic) philosophical lexicon generally names *problem of individuation* the question about what makes of a certain object the object it is. For example, we can ask what makes of this pen *exactly this pen*. *Individuation* has thus to provide us with a criterion describing what must be the case for something to be *the same thing* as that object [cf. Recanati 1993, p. 170]. But the edges of this task are rather faded and controversial. Dretske, for example, contrary to Geach and Wiggins, has argued that one *can* refer to an object, or think of an object, without knowing what sort of object it is [cf. ibid. and Dretske 1988, p. 73]. Part of the problem is subject matter of classic analytic metaphysics, part of cognitive psychology. Recanati, referring to the work of Treisman [1992], outlines two distinct levels of identification which are involved in perception: the localization of an object as space-occupier, and its identification as a certain type of object (a plane, a bird, or a black horse) [cf. Recanati 1993, p. 171].

My idea of individuation is rather between these two levels, that is, after pure spatial (and eventually temporal) localization, but prior to kind-identification. I refer to a cognitive process as *individuation* of an artifact when, for example, one, ignoring what is the object made up of metal wings and a screw presented to her/him, is anyway able to recognize it as an artifact. An individual finally identifies that item when (s)he is able to recognize it as a corkscrew.

This distinction is plausible from an intuitive and conceptual point of view, and seems to be acceptable from a logical as well as an epistemic point of view. There are many artifacts whose meaning and function is ignored. Field archeologists often find objects, which they recognize as clearly human-made, even though they are not able to identify them. Analogous samples of unrecognized artifacts have triggered the fancy of the most imaginative people, like in the case of the *Coso Artifact*. The *Coso Artifact* is a cylinder of a white material looking like porcelain, of the size of a few centimeters, contained in a geode that some amateur geologists found nearby Olancha, California, in 1961. This finding gave rise to many conjectures about the origin of this curious item[2], because a geode is a rocky nodule with fossil origins, that sometimes take several hundred thousands years to develop.

The discoverers of the *Coso Artifact*, had available for recognition only a certain amount of information concerning its material and structure, and nothing else. However, they recognized it, maybe with

2 Cfr. <http://www.talkorigins.org/faqs/coso.html>.

some approximation, as an artifact. Thus, was the information available, enough for the task that I have named of *individuation*? Are, generally speaking, matter and shape the only two aspects required in order to individuate an artifact?

3 Understanding Artifacts

The issue at stake regards what does drive an individual to perceive as an artifact an object whose history and function is ignored, and of which one knows only its material composition and structural configuration.

Several years ago the biologist and philosopher Jacques Monod sketched a possible solution for this question [*Le Hazard et la Necessité*, 1971]. According to Monod the regularity and geometry of shapes is the only possible symptom of *artifactuality*. Such a solution is not very satisfactory and rather restrictive. Of course the regular and geometric forms of an object lead to higher probabilities that someone will recognize it as an artifact, with respect to an object that does not display these forms. However the meaning of these probabilities has to be explained: it is not plausible that we have developed a habit to recognize as artifacts those objects with regular and geometric shapes, only by abduction from a relatively limited number of empirical cases which we find in our experience. Moreover, there can be artifacts, which do not display any regularity or geometrical design (such as certain cookies or many other items which are intentionally made irregular), as well as there may be natural items which instead display these features (like a shell, for example).

Other hypotheses can pivot specifically on how our cognitive system works. During the last years a general strategy has gained ground, which consists of explaining all natural phenomena, inclusive of the human beings and of their more or less complex behavior, within the theoretical framework of the Darwinian theory of evolution (Neo-Darwinism). The recent program of evolutionary psychology has developed reasonable explanations of many mental faculties. There can be an evolutionary account also of many capacities which allow us to understand the most important features of the world on the basis of our common sense (and of which, therefore, all normal human beings are endowed). Leda Cosmides and John Tooby, two of the main champions of Neo-Darwinism, for example, argue for a plurality of domains consisting of '*dedicated intelligence*', that is

> … rapid learning guided by specialized domains of core knowledge, which allow perception of, attention to, and reasoning about important classes of entities in the children's environment (such as number, objects mechanics and agency) [Defeyter and German 2003, p. 135].

If we adopt this theoretical framework, we could hypothesize that human beings are endowed with a special mechanism, or better, a cognitive module or *organ*, which is finalized to the individuation of artifacts. There are many supposed specialized mental modules, having different particular functions, such as to detect the direction of one's glance, to recognize the targets of an intentional agent, even to *reason* about the mental states of other people.

I am rather skeptical about the plausibility of this hypothesis. I think that, at a phylogenetic level, the adaptation of an organism takes a long time, and concerns more general features. Therefore I consider bizarre the idea that our brain has been shaped in a way such that it can recognize a so big and multifarious class of objects, which have appeared only in the recent past of the history of our planet. Natural elements are an integral part of the surrounding environment, and they have an evolutionary time similar to ours; artifacts instead are subject to much more frenetic evolutionary rhythms. A human being, lived in Europe just one century ago was not particularly different from a contemporary human being. Instead, the environment in which a contemporary human being lives appears as considerably different from that one of her/his ancestors, one century ago. A salient aspect for which the two environments differ consists exactly of the different features of artifacts. Even though the artifacts of one century ago are artifacts like those in our contemporary world. Probably it is matter of different classes of artifacts, but they all belong to the general category of artifacts. A coal or a wood-burning stove is different from a modern electric or gas stove, but they are all heating systems, like a candle or an oil lamp are different from an electric lamp, even though they are all lighting systems. And they are all artifacts. Thus, if our cognitive system have adapted in way such to develop a specific module for the individuation of artifacts, it should be based on a criterion of recognition relatively simple and precise; but this keeps being unknown. The reason of this is maybe because such a module does not exist, like it is likely that a capacity for individuating artifacts, if by it one means a specific mental faculty, does not exist either.

There is another more plausible hypothesis, yet coherent with the theoretical framework of evolutionary psychology. It also postulates specific domains of knowledge, which likely would be in charge of more elementary mental activities. They could be functional to the elaboration of more basic information processing tasks, like the representation of the physical properties of an object, from which one could derive the range of motor activities that can be implemented with it. For example, one by watching a Lego tile could infer its flexibility, its rigidity, its extensibility, etc., as well as a number of possible interaction that it can play with other objects or parts of an object, such as its modularity, its compositionality, its adhesiveness. Modular competencies of this kind could be functional for the understanding of what Gibson [1979] has named as the *affordances* of an object, that is, the actions which can be performed by means of it. Through a sequence of exceptionally complex computations, which elaborate the possible combinations of the affordances of the different aspects of an object, one could reach a clear synoptic picture. This could synthesize in a unique representation all the features, which determine an artifact, and eventually also its function.

I am skeptical also about this hypothesis. First of all it requires a too complex cognitive system in charge of the processing of the computations, which end up with artifact concepts. It is difficult to understand how it could avoid the risk of computational collapse and/or explosion. Moreover, the idea that a certain object expresses by means of the display of its affordances its own general artifactual feature does not define its individuation, but rather its identification. The affordances of, e.g. a hammer (I take into consideration its most salient properties), are expressed, at a certain level, by its handle and by the metal ingot in which it is stick. The handle displays the possibility of the hammer to be easily grasped in a certain way, and the ingot displays the suitability of the hammer for percussion. This is already a cognitive task, which aims at identifying, rather that individuating, the artifact at stake. Probably a hammer is a tool so popular that the idea of distinguishing its being a member of a certain class of artifacts from the idea of its being a member of the general category of artifacts sounds rather odd. The *Coso Artifact*, again, does not express any specific affordance, yet it is (and it was) perceived as an artifact, even though none was able to understand *what kind of artifact*.

4 The Interpretation of Artifacts

From these considerations one draws a conclusion that hardly can be disputed: the individuation of artifacts, as it is an immediate process, is not a simple and primitive cognitive procedure. I think that it requires, also at such an elementary level, a plurality of cognitive resources, whose activities converge in a form of synthetic judgement that makes of the recognition (individuation) of an artifact an aware process. It is therefore a process different from more basic cognitive processes, which are often unaware (like the way we give meaning to images). At a phenomenal level we can individuate artifacts with a good level of approximation, but with no certainty. The mode in which we individuate artifacts seems to be a more articulated form of recognition that appears to be more complex than the mode in which we recognize an animal, a vegetable, or a stone. Rather than a specific cognitive faculty, or a cluster of specialized cognitive faculties, our capacity to individuate artifacts appears to be based on a heuristic procedure, which is rooted in our experience and has a fairly robust normative character. As such, it has to be considered a form of primitive, *radical interpretation*. This kind of interpretative procedure is particular, because is prior to any hermeneutic task based on aware beliefs and desire, and starts with perceptual cues, which are grasped in specific aspects of an object and of the way it is presented.

One of the main problems in the understanding of this complex cognitive procedure is the definition of these particular perceptual cues. According to a previous hypothesis of mine, I had identified them with elementary *indexes for action*. I described these as certain intrinsic characteristics of an object (like shape, size, etc.), which are codified by our cognitive system in relation to the kind of action (such as, e.g. prehension) that can be performed by means of the object at stake. In other words, *indexes for action* were conceived as a peculiar kind of *affordances* detectable in all kinds of artifacts. However, these indexes can be recognized in all the objects which have suitable dimensions: both the *Coso Artifact* and a stone of analogous size can be codified in terms of the same kind of actions, for example a catch. The *indexes* are effective in prompting very general kinds of actions, they do not help much in understanding the artifactual nature of an object, even though they may contribute in restricting the range of possible way in which that object can be used. Also geometry and regularity of shapes, as well as its material, can help us in understanding that an object is an artifact, as suggested by Monod. But these elements cannot, by themselves alone, trigger our process of individuation. Let's suppose that we find within a rocky nodule, within a geode like the one of the *Coso Artifact*, an unknown crystalline material (crystals usually have a fairly regular polyhedral form). Rather than thinking about the finding of an artifact, one would likely be thinking of a novel kind of crystalline material, which is present in nature but unknown to us.

5 The Capacity of Recognizing Artifacts

The individuation of an artifact seems rather an interpretative procedure. As a (special) form of interpretation, it depends on the whole and aware cognitive perspective of the subject who performs it, and on her/his background attitude relative to this task. This is the cognitive space that an agent establishes with the object of its interpretation. Such a space is rather a form of cognitive dimension required by an individual in order for him to project in it the item that (s)he is about to interpret as an artifact. It is this cognitive space itself that is a presupposition for the possibility to recognize something as an artifact, and then, as an artifact of a certain kind. Without postulating the cognitive dimension in which to localize something as an artifact, the artifact itself could not be conceivable.

In other words, I think that one's capacity of individuating an artifact presupposes the capacity of defining a certain *cognitive* dimension where to place the item that is going to be recognized as an artifact. This dimension is not *spatial* (in a physical connotation of this term). Thus we have to understand which is its nature. My (tentative) answer is based on the primary understanding of *artifacts as means to reach ends*. The cognitive space in which we understand artifacts, is first of all a space defined by a system of axes and coordinates, which correspond to means and ends. It is a dimension built on a basic mental frame that consists of a teleo-instrumental reasoning. This frame is intrinsic of the way in which human beings reason in relation to some phenomena, regarding mainly the context of action, agency, and intentionality [in a non-technical, non-Searlian meaning; cf. Searle 1992]. As we apply the *means-end* psychological pattern to the domain of intentional behavior, so we do in the case of artifacts. And the reason is consequential: artifacts are products of intentional action performed conforming to the dynamics of the means-end reasoning. This dynamics rules our practical life, and it reflects our way (the unique way we know) of thinking about the achievement of something, that is, *by means* of something else. Therefore, all the possible symptoms of artifactuality which we ascribe to this or that object, are rooted in a certain attitude that directs our understanding towards it as a created by a human mind according to the means-end reasoning.

Within this perspective, the general idea of an artifact of which we do not know the identity does make sense. It does because we think of the artifact as placed in the heuristic/pragmatic space that depicts it as some means aimed at pursuing some end. The projection of the means-end frame is in a certain sense *a priori*, with respect to the recognition (individuation and identification) of the artifact itself. Once the teleo-instrumental mode of reasoning is on-line, each element on which it ranges over is potentially subject to be interpreted as an artifact. The projection of this teleo-instrumental modality of thought is by default, even though it may depend on certain plausible standards. Applying this default mode of reasoning should not clash with empirical evidence and with other tested explanations (such as causal explanations). The teleo-instrumental frame, in fact, has been used in the past by *default* without respect for these conditions. Scholastic philosophers, for example, have tried to use it in order to prove the existence of God. Other (well respected) philosophers (like Hume, Kant, etc.) had not trouble in rejecting its wrong employment in this case.

The perception of a certain environment through the filter of the teleo-instrumental explanatory scheme gives a meaning (within the context of the means-end reasoning) to all the elements on which it projects its logic by default. Only to a few of these elements it is possible to ascribe a definite meaning; these are the artifacts, which we can provide with a coherent explanation. Other items, instead, would make no sense if interpreted according to the teleo-instrumental dynamics. Yet other objects can be endowed only with a partial, and therefore indeterminate, interpretation. These are the artifacts whose identity is ignored but which are individuated as artifacts.

6 A Brief Conclusion

These are the fundamental theses, which ground my intuitions: 1) the concept of artifact is conceivable only as the result of the projection of a teleo-instrumental explanatory, heuristic, and pragmatic mental pattern; 2) this explanatory teleo-instrumental framework is projected by default on the basis of the means-end reasoning that each of us use in a practical dimension. They are not particularly original theses.

Immanuel Kant, in the *Critique of the Faculty of Judgment*, appealed to the notion of *purposiveness without a purpose* which is near to the idea of an artifact without identity that I have taken into consideration. More recently Daniel Dennett, suggested an explanation of the human understanding of artifacts that was based on the 'Design Stance' that is similar to my second thesis. Of course there are also significant differences. According to Kant the essence of teleology was a regulative ideal. I think instead that the teleo-instrumental frame that I have sketched is a real heuristic criterion, efficacious and effective, driving our cognitive system. Dennett believes that the 'Design Stance' is only a cognitive trick, a shortcut explanatory strategy that can be used also in order to explain natural phenomena. I think instead that it is not only an explanatory tool, but also a concrete and operational scheme for the planning of all human actions.

Along my *argumentation* there are many points left without an adequate explanation. The main idea of a teleo-instrumental explanatory frame has to be justified with solid arguments. I don't think that it is only an *a priori* presupposition in a logical sense. I think it is also a concrete structure of our cognitive system. As such, this subject matter is worth to be studied with the full support of those cognitive sciences, which are relevant or can be relevant for a progress of its understanding.

Bibliography

W. Ahn, 'Why are different features central for natural kinds and artifacts?: the role of causal status in determining feature centrality', in *Cognition*, 69, 1998, pp. 135-178.

I. Biederman, 'Recognition by components: a theory of human image understanding', in *Psychological Review*, 94, 1987, pp. 115-147.

P. Bloom, 'Intention, history, and artifact concepts', in *Cognition*, 60, 1996, pp. 1-29.

S. Gelman and P. Bloom, 'Young children are sentitive to how an object was created when deciding what to name it', in *Cognition*, 76, 2000, pp. 91-103.

A. Defeyter and T. German, 'Acquiring an understanding of design: evidence from children's insight problem solving', in *Cognition*, 89, 2003, pp. 133-155.

D. Dennett, *The Intentional Stance*, MIT Press, Cambridge MA 1987.

D. Dennett, 'The interpretation of texts, people, and other artifacts', in *Philosophy and Phenomenological Research*, 50, 1990, pp. 177-194.

F. Dretske, *The Explanation of Behavior*, MIT Press, Oxford MA 1988.

J. J. Gibson, *The Ecological Approach to Visual Perception*, Houghton Mifflin, New York 1979.

B. Landau, L. Smith, and S. Jones, 'Object shape, object function, and object name', in *Journal of Memory and Language*, 38, 1998, pp. 1-27.

B. Z. Mahon and A. Caramazza, 'The organization of conceptual knowledge in the brain: living kinds and artifacts', in E. Margolis, and S. Lawrence (eds.), *Creations of the Mind: Essays on Artifacts and their Representations*, OUP, Oxford (to be published).

A. Martin, L. G. Ungerleider, and J. V. Haxby, 'Category specificity and the brain: the sensory-motor model of semantic representations of objects', in M. S. Gazzaniga (ed.), *The Cognitive Neurosciences* (2nd Edition), MIT Press, Cambridge MA 2001, pp. 1023-1036.

A. Matan and S. Carey, 'Developmental changes within the core of artifact concepts', in *Cognition*, 78, 2001, pp. 1-26.

J. Monod, *Le hazard et la necessité*, Edition de Seuil, Paris 1971.

F. Recanati, *Direct Reference: From Language to Thought*, Blackwell, Oxford 1993.

J. R. Searle, *Intentionality*, CUP, Cambridge 1983.

A. Treisman, 'Perceiving and re-perceiving objects', in *American Psychologist*, 47, 1992, pp. 862-875.

Giacomo Romano, Department of Technology Management, Eindhoven University of Technology

David Smith

Fly's Eye View: an investigation of the impact of the experience of using a VR system on young children's understanding of representation and reality

Abstract:

This paper presents a case-study based on an attempt to design, develop and demonstrate an experimental approach to the initiation of small children into some founding ideas about digital art, but which also taught them something about themselves. The chapter sets out and explains some of the challenges, problems and compromises involved. It describes 'Fly's Eye View', a project devised and directed by David Smith and his team, and supported by a grant from the UK Techniquest/Pantechnicon Millennium Awards Scheme, which were intended to help bridge the gap in public perception between science and art. 'Fly's Eye View' was specifically aimed at young children, aged 7-11 years. The project used an artifical environment, a low-level VR fly-by system, to represent the world as it might be experienced by a dragonfly.

The analogy between the mosaic visual 'perception' of insects and image representation in computer systems allows some interesting explorations into what may be involved in 'seeing'. This analogy is not perfect, however, and could easily be misleading if taken too far. However, it was used as the starting point for a series of exploratory activities, using digital artwork to investigate questions of experience, reality and representation.

The paper reviews the outcomes of the 'Fly's Eye View' project and discusses their possible significance in the context of approaches to art education and the stimulation of conscious reflexivity through the experience of using an artificial environment.

Keywords:
children, digital art, virtual reality (VR), representation, consciousness, the artificial

1 Introduction

According to neo-Darwinian models of evolution, the fate of species is not determined by effects at the level of the individual, but is rather a function of continuous variation at the population level. In nature, the main sources of variation are the genetic mechanisms of mutation and recombination. But through the agency of a range of cultural factors, the human species is becoming increasingly independent of the purely genetic factors which determine the path of evolution in other species. Nevertheless, it can be argued that in an era of 'post-biological' evolution, variation and diversity will still assume as much evolutionary significance as before. It is, rather, the dynamics of evolution that will change. Because science has increasing power to permit us to intervene in the biological processes of evolution, and thus to minimise their impact, human creativity now provides one of the most fertile sources of extragenetic diversity. Art and artistic thinking are vitally important factors in human creativity, and hence in the generation of diversity. Artists constantly invent new ways of seeing and new forms of behaviour which may remain esoteric, but which may also achieve the status of 'memes': words, ideas, concepts or fashions that spread rapidly through a culture [Dawkins, 1989].

The way in which the next generation is brought into contact with 'artistic thinking', rather than with a canonical corpus of art products, therefore assumes a much greater importance than contemporary society often concedes. Formal education and other structured formative experiences may be seen as non-genetic ('artificial') evolutionary mechanisms maintaining and, ultimately, generating species diversity. In a situation in which the creative consciousness has already assumed great evolutionary significance (for example, through the development of modern scientific medicine), it becomes vitally important that education should aim to initiate young people into the processes by which cultural diversity is created, rather than simply to transmit received forms.

The possibility of using computers to represent different ways of looking at the world, and their potential as tools for the creation of quite new artistic media in their own right, have excited artists since the sixties. The new forms have both threatened and extended current practice. Young people today are growing up alongside new interactive media which are able to influence interpersonal communication by shifting perceptions of both what it is possible to say and how it may be said. It is entirely probable that they are unconsciously participating in the creation of new artificial symbolic forms ('languages') concerned with sequencing and association of images, sounds and conventional text. Here is the basis for an entirely new and quite unprecedented source of artificial variation, with the added possibility of near-instantaneous transmission.

Societies have always transmitted fundamental cultural information in forms such as dance, music, myths, drama and iconic symbolism. Multimedia technologies offer the capability for expression in a complex fusion of verbal and non-verbal forms, and we can only guess what the outcomes of this process might eventually be. We could witness a rediscovery of these modes of communication as everyday forms, rather than high 'arts'. But for this to happen, we need to give attention to children's learning experiences, and here we seem to be failing.

At present, educators and politicians seem equally focused on the short-term audit of educational treatments, rather than on the liberating possibilities of the learning process. Education in the arts is often marginalised: seen as a (barely necessary) cultural gloss on the real business of preparing young people for the 'world of work'. There is little room for playful experimentation or unorthodox methods. All too much of the educational system is concerned with control, rather than development, and we seem to have reverted, in less than a generation, to the thinking of Athens in the 5th Century BC, where Plato was quite emphatic about the need to regulate and censor children's learning:

> … we soon reap the fruits of literature in life, and prolonged indulgence in any form of literature leaves its mark on the moral nature of a man, affecting not only the mind but physical poise and intonation … [*Republic*, p. 141.]

Plato, setting the tone for many of our own contemporaries, saw art and artists as potential threats to his vision of a perfectly ordered society:

> It is not only to the poets therefore that we must issue orders requiring them to represent good character in their poems or not to write at all: we must issue similar orders to all artists and prevent them portraying bad character, ill-discipline, meanness or ugliness in painting, sculpture, architecture or any work of art, and if they are unable to comply, they must be forbidden to practise their art .... Our artists and craftsmen must be capable of perceiving the real nature of what is beautiful, and then our young men ... will benefit because all the works of art they see and hear influence them for good ... insensibly moulding them into sympathy and conformity with what is rational and right. [*Ibid.*, p. 142.]

But we cannot attempt to bring up children in an environment of restriction, even repression, and still expect them to be courageous creative thinkers on the scale which will be rendered necessary by the ecological and economic disasters which seem to be only just around the corner. Perhaps the great Japanese poet Basho had a clearer view of what artistic creativity really involves when he wrote:

> With a little madness in me
> Which is poetry
> I plod along like Chikusai
> Among the wails of the wind

To Basho and many of his associates, the creative mind operated in a plane of primary consciousness which was qualitatively different from the analytic and mechanistic. Perhaps we could assert that an educational process for art which does not at least encourage access to this 'little madness' seems inevitably doomed to failure. There is, however, an additional, and perhaps more significant dimension to this issue, which may be illustrated anecdotally.

Many years ago now, I took my son to see one of the 'Superman' films, showing locally to a largely teenage audience. The story had just reached the 'romantic interest' point. Superman was kissing Lois Lane! However, instead of either waiting impatiently for the action to be resumed or else wallowing in the pleasures of vicarious lust, the audience started chanting 'Fast Forward! Fast Forward! Fast Forward!' The young people in the audience were both 'film-literate' and sophisticated enough to derive an appropriate metaphor from their familiarity with video-recording technology. But there was something else, too: something which is germane to the present discussion.

The young audience were not merely receiving a text in the passive sense so beloved of many uninformed critics of the media. There was an implicit expectation of control, in which they were sharing a perception of themselves as having an interactive relationship with technology and technological artefacts in a more complex sense than simply in terms of 'human-machine interaction'. This perception is now one of the cultural realities of our society, and we can expect it to grow dramatically.

If children appear to have intuitive relationships with the technological artefacts which increasingly define the parameters of their world, these relationships offer little more than an impoverished perspective if they do not transcend the mere details of control (though this is certainly important). There is real danger that the ability of the computer to create exact copies may predominate over our willingness and ability to use the technology to distort, reconfigure and create. By permitting the setting of arbitrary standards of output, the conventions of precision and exact representation may come to be used, even unintentionally [see Sinding-Larsen 1991], as disciplinary and controlling devices, just as Papert [Papert, 1980] argued that the rules of formal reasoning were used to suppress the use of intuition. It seems to us vital that both teachers and children be able to experience the creative and cognitive affordances of computing technology in terms other than of data processing and replication.

There seem to be many advantages in offering school students an opportunity to experience some of the exciting possibilities opened up by creative use of computers and other digital resources as tools for the manipulation of images. The ubiquity of computer-based technology and the extending phenomenon of the internet are at the forefront of issues of identity and persona which are part of what might be called the 'demotic philosophy' of communication space. That is to say, the way in which ordinary people understand and reason, as opposed to the formal analyses of professional literati. Coming to terms with these issues is likely to be a necessary (if not sufficient) precondition of living, creating and possibly even remaining sane in the chaotic world of what Massimo Negrotti [Negrotti, 1991] has described as the 'culture of the artificial'.

Gigliotti [Gigliotti, 1998] further surmises that one purpose of consciousness is to help us make our way through constant change. Questions of consciousness, and the ideas and lines of reasoning which flow from them, are therefore far too important to be left to philosophers. They surely become proper subjects for education, and the creative arts seem, *prima facie*, to be an appropriate locus for their exploration.

2 Fly's Eye View: The Project

Although I am a scientist by education, most of my work over the past few years has been concerned with the creative application of new technologies. I therefore wanted young people to be able to experience the excitement and fascination of the interface between art and science, and I believed that one way of doing this was through the unique opportunities which computer-based virtual reality (VR) offered for exploring the field of visual perception.

The centrepiece of the work was a 'virtual reality dragonfly'. (There was no specific reason for choosing a dragonfly: it might just as easily have been a cockroach or a bedbug!) A low-footprint PC computer was installed inside a large (i.e. 4.5 metre wingspan and 4 metre body-length)[1] dragonfly. The 'fly' was conceived as a sculptural realisation: that is, as an artwork in itself, in order to reinforce the artistic context of the project. Children aged between 7 and 12 (or *very* small adults!) were able to 'fly' over a garden scene, comparing the visual world of a bird or a human with that of a simulated dragonfly. Associated contextualising activities included:

- Group observation drawing/painting activities to create an exhibition for display at the chosen site.

1 Possibly the largest VR helmet in history!

- Computer graphics workshops

'Fly's Eye View' set out to explore questions of perception and representation, using a form of 'virtual reality' experience comparing insect and human vision, and a range of investigative learning activities based on and around digital art. Using an (admittedly questionable) analogy between the visual 'perception'[2] of insects and image representation in computer systems allowed some interesting exploration into what might be involved in 'seeing'. One issue which we had to tackle concerned the level of 'realism' required in our simulation. Our original plan was to provide children with the possibility of switching between an accurate 'mosaic' view and a conventional 'picture' view. However, we concluded that the mosaic image would give rise to unnecessary complexities of interpretation, since a physiologically exact representation would beg important questions about what was actually perceived by an insect. It seemed more constructive to pose these issues as problems and to explore them with the children, who were, indeed, often very intrigued by the thought that we cannot really be certain about any perceptions – even our own. We therefore made no attempt to provide an exact representation of the insect's mosaic vision. Instead, we used what Brenda Laurel [Laurel, 1993] described as 'viewpoint-dependent imaging', constructing the imagery from the point of view of a person (or dragonfly) located in the virtual environment. We were, after all, far more interested in mediating reflection on subjective experience rather than in entomology per se!

The analogy between insect and computer vision is not a perfect one, and could even be misleading if taken too far. We were certainly aware of this problem, and tried not to overstate the relationship, which was the conceptual starting point for a series of exploratory activities, using digital art work to investigate questions of experience, reality and representation. Worksheets and teacher-led activities invited children to speculate about what 'reality' is, and how it might be represented and distorted, and to consider such questions as:

2 Used rather loosely here – we were dealing with young children, and we were not teaching philosophy!

- How can digital art help us to understand different ways of seeing the world?
- How can different ways of seeing the world help us to understand digital art?

We wished to offer young children some early insights into possible new art forms and ways of looking at the world around them: quite apart from giving them an opportunity to learn about some up-to-date science of visual perception. The VR display allowed direct comparisons to be made, whilst worksheets and other activities directed attention to digital art and to wider issues of perception (at a level commensurate with the children's levels of cognitive development).

The project was launched in Newport, south-east Wales. It then travelled to a variety of locations throughout Wales, including the National Museum and Art Gallery in Cardiff. The final event of the 'Fly's Eye View' tour was the National Eisteddfod: an important annual festival dedicated to the Welsh language and culture. Each location (except the Eisteddfod) involved two weeks of on-site work. Approximately 2000 children were involved in the project.

A typical working session began not with the computer but with 'warm-up' pen-and-paper drawing exercises, such as sketching with the eyes closed. Children (and teachers) enjoyed this very much, and it was very effective in focusing attention onto specific tasks and ideas, rather than onto technology per se. This was particularly important in view of the sheer size of the 'dragonfly', which was designed to dominate every space in which it was placed, and the obvious presence of the computers (obsolescent Macintoshes lent by the university's Multi-Arts Computer Centre), together with low-cost colour printers and scanners[3].

To facilitate supervision, the children were than divided into two groups. One group were introduced to the computer-based art task and then supervised by their own teachers. They focused on making and modifying digital images. They worked with digital still cameras to produce group images, which were then loaded onto the computers

3 A full inventory can be supplied on request.

and manipulated with simple low-cost drawing tools[4]. The object here was to reassure teachers that significant work could be undertaken on 'ordinary' computers and with simple low-cost or 'freeware' programs. The original images were self-posed and photographed, and then developed to a theme selected by the children. There was a wide variety of themes among the girls, whilst the boys' taste tended to run to monsters!

A number of points emerged from this phase of the session:

- Very few of the children experienced any major problems with digital photography or with any of the software;
- Most of the more computer-phobic teachers soon gained confidence, and were able to participate fully;
- These activities were also within the capabilities of a group of adults with severe learning difficulties who attended a session in the Welsh National Museum and Art Gallery, Cardiff;
- The creation and manipulation of digital pictures helped children to think of 'photographic' images as starting points for creativity, rather than as archival end-points;
- Most of the children were able to understand a simple account of how computer-based images were created, and many were very intrigued by the idea of being able to edit down to the level of the individual pixel.

Whilst this was in progress, another group of children were focused on activities around the dragonfly itself. Individuals had a chance to be a dragonfly and to experience 'flying' round a garden. Time pressures meant that this experience was necessarily rather short – although some of the children were extremely difficult to extract from the system! The original plan had been to use an advanced 'force-effect' joystick, but pilot tests convinced us that this created an ambience which was too close to that of the *shoot'em up* game. Flight control throughout the project was through a normal mouse rolling on a modified proprietary 'telephone arm'. This arrangement caused no problems, and all of the children – and most of the adults – found the mouse easy to use.

4 Either 'Color-It' version 2.3 or a version of 'Photoshop Lite' which had been bundled with the computers and digital cameras (no special significance was attached to the choice of image software).

When the groups were eventually combined, the children were told about an insect's mosaic vision, and this was analogised to the pixellation of the computer screen image. The children were asked to imagine what a dragonfly might *really* be seeing, as opposed to what they had seen in the VR simulation. Many, though not all, of the children were able to grasp that they had no idea what a member of another species, or for that matter, another child in their own class, perceived as 'reality'. It seemed that the children were generally able to take on board both the Cartesian '*cogito*' and the age-old question of artistic reality. This led to some lively discussions!

Posing questions about the nature of perception had the effect of dislocating the children's primary experience as 'users' of a computer-based representation system, and relocating them in another sphere of conscious activity. What was taken for granted as 'simple' experience was restated as a problematic issue. This is analogous to the moment of consciousness implicit in Heidegger's conception of a tool's transition from the state of being 'ready-to-hand' to its being 'present-at-hand' when the primary experience of its use is disrupted.

Follow-up studies were carried out in the schools. An evaluation questionnaire was designed to assess children's and teachers' perceptions of their experience, and results from this survey show very high levels of satisfaction. Indeed, one of the few pointed criticisms (by a boy!) concerned the lack of things to kill in 'the game'. We have yet to establish the stability of the conceptual understanding apparently demonstrated by at least some of the children, and funds are now being sought to study this particular aspect of the project in a more systematic way. All in all, 'Fly's Eye View' appears to have touched on a very fruitful area for creative computer-based education, which will bear further development.

3 Conclusion

Despite its growing importance as a field or contemporary art practice, it is still quite rare for computer-based digital art to be experienced by school pupils. We believe that 'Fly's Eye View' has been successful even at this limited level, in that we have been able to demonstrate ways in which quite basic computers can provide foundations for exciting creative practices. Much more importantly, however, we have found that young children have the capacity to be intrigued and excited by some very complex ideas and philosophical issues. We do not, assert that this will be will necessarily result in the production of a generation of young philosophers. We are perhaps more inclined to accept O'Nualláin's [O'Nualláin, 1994] argument that consciousness is best thought of as an occasional achievement in mental life, rather than a continual process.

We have at least demonstrated that the awakening of 'metaconsciousness', as an awareness of issues of consciousness and being, can be stimulated, even in young children, by appropriate educational environments. Although conventional wisdom might question whether this could or even should be achieved, we have accepted the assertion of Vygotsky [Vygotsky, 1978] that the only good learning is that which is in advance of development. Following Papert [Papert, 1980], we designed a post-Piagetian 'constructivist' learning experience which could be emulated by 'ordinary' teachers. We were able to demonstrate that even young children are able to respond to the ideas underlying the project, and that some are able to speculate in an interesting way and even to begin to construct quite promising digital art on the basis of that speculation. Perhaps this illustrates above all a mechanism in which the transition from one level of user experience to another facilitates what Vygotsky [*ibid.*, 1978] called 'the zone of proximal development'. We have also built upon Papert's concept of 'powerful ideas' as intellectual tools – in this case, conscious reflection on the imagined experience of an entirely different organism – but we have also implicitly located these issues in terms of our relationships with the artificial:

> … how can we hope to understand and develop a positive relationship with beings of our devising if we understand so little of the incredible richness of those beings that already exist and share our conscious and unconscious space here and now …? [Gigliotti, *ibid.* p. 37]

The 'dragonfly' theme provided a valuable (but not essential) vehicle for this work. We believe that, flawed though the approach may be, quasi-empirical studies such as that described here are essential to making progress in our thinking in this important domain.

Bibliography

M. Basho, 'The Records of a weather-exposed skeleton', in *The Narrow Road to the Deep North* (translated by N. Yuasa), Penguin, London 1966.

R. Dawkins, *The Selfish Gene*, OUP, Oxford 1989.

C. Gigliotti, 'What is consciousness for?', in *Digital Creativity*, 9, 1, 1998, pp. 33-37.

M. Heidegger, *Being and Time* (translated by J. Maquarrie and E. Robinson), Blackwell, Oxford 1962.

B. Laurel, *Computers as Theatre*, Addison Wesley, New York 1993, p. 204.

M. Negrotti (ed.), *Understanding the Artificial*, Springer-Verlag, London 1991.

M. Negrotti, *The Theory of the Artificial*, Intellect Books, Exeter 1999.

S. O'Nualláin, 'Some consequences of current treatments of consciousness and selfhood', in *AI & Society*, 8, 1994, 4, pp. 305-314.

S. Papert, *Mindstorms: Children, Computers and Powerful Ideas*, Harvester, Brighton 1980.

Plato, *The Republic* (translated by H. Lee), Penguin, London 1955.

H. Sinding-Larsen, 'Computers, musical notation and the externalisation of knowledge', in M. Negrotti (ed.), *Understanding the Artificial*, Springer-Verlag, London 1991, pp. 101-125.

L. S. Vygotsky, *Mind in Society* (edited by M. Cole, V. John-Steiner, S. Scribner and E. Souberman), Harvard University Press, Cambridge MA 1978.

David Smith, School of Art, Media and Design, University of Wales College, UK

Mariselda Tessarolo

Transition of the relation between artist and user to that between culture supply and demand

Abstract:

It is impossible for the public and the cultural product to meet unless the role of culture in society is understood and the characteristics of the organizations in which culture is diffused are examined. Cultural products cannot be understood separately from the settings in which they are produced or consumed [Crane, 1997]. The relationship between the artist and the user has never been simple. Suffice to think that such relationship involves as many as three disciplinary contexts: sociology of the artist, of the work of art (creation, production, purchase) and of fruition (and/or consumption). Aesthetic form and content have always been separable, and so capable of being de-contextualized and handled separately: art as either a real or symbolic object is only the starting point, the organizational center for the analysis of the social processes. In this way the intellectual import of the sociology of art is shown to be very high [Zolberg, 1994].

Nowadays, what makes the cultural object particular is its belonging to a sector, its fragmentation, its belonging, despite everything, to a globalised world. The cultural object is studied, commissioned and produced to be enjoyed and appreciated by the public at large, to be inserted in a network in which it is distributed through information and use. Telecommunication instruments and new technologies have shaped new forms that are becoming increasingly familiar and therefore desirable. Because of this, fruition has changed its appearance of individuality to assume a semi-collective guise more fitting to its user.

The game of homogenization and differentiation is indeed born from the hyper-segmentation of issues and life styles. Such a game guarantees creators of culture a great control over their products. The category of 'cultural products' has undergone a widening that has led users to choose niches of specialized consumption. Paradoxically, users meet because they have similar restricted preferences. There is a dynamics between supply of culture and demand for it, in which success uncertainty is high [Griswold, 1997]. Indeed, final success mainly depends on the public, who actually has the last word. Consumption complexity is sometimes unpredictable. This shows that the user is not only a knot in the net, but rather someone that has to be convinced and then inserted in the complex consumption process. Therefore, the user can influence production and, because of this, innovation has to be predictable. The continual input between communication, production and consumption causes circularity, promiscuity and changes that help socialization and so the acceptance of the product and of the organization that implements it.

Keywords:
distribution of culture, cultural industries, supply and demand for culture

1 Premise

Dynamism, necessary to the existence of the social system, clashes with the static equilibrium needed to determine which rules must become norms: and so two forces – centripetal and centrifugal – are at odds. The social scientist experiences the difficulty of constructing ideal models that do not correspond to empirical ones, which are in any case characterized by dynamism.

Sensible actions and rationality are not a once-and-for-all thing; they are constructed socially and are equivalent to actions actually present in everyday life. The social actors and the system are in continuous evolution, since society itself is a set of relationships, each of which transforms the terms that it connects[1].

Institutionalization has intervened in every age, as society has always had to 'check' human creativity, on the one side, and to 'protect' it on the other. Only in this way has art become a 'social fact' and obtained legitimization in the institutionalized channels implemented by the selective processes operated by the 'gatekeepers', that is to say representatives of a given culture. Their task is to check the 'purity' of a model and the 'conformity' of new creations to existing models.

The artist-user relationship has never been easy. It suffice to think that it involves as many as three different disciplinary settings: the so-

1 Piaget observes that the relationship between subject and material object changes both the subject and the object, because of the assimilation of the latter to the former and the adaptation of the former to the latter [1950]. Changes take place also because of the interaction between subjects: in the interaction the subject changes.

ciology of the artist, of the work of art (creation, production, assignment), and of fruition (and/or consumption). However, aesthetic form and content are separable, therefore capable of being decontextualized and manipulated separately (each on its own): art as a real or symbolic object is a starting point or, in any case, the organization center for the analysis of the social processes. This highlights the intellectual import of sociology of art, of cultural goods and of their consumption.

Contact between the public and the cultural product is impossible, unless the role of culture in society is understood and unless the characteristics of the organizations within which and from which culture itself is spread are examined. Cultural products, in fact, cannot be understood separately from the contexts in which they are produced or consumed.

Nowadays, what makes the cultural object particular is its belonging to several sectors and its fragmentation. The cultural object is studied assigned and produced to be enjoyed and appreciated by the public at large and to be conveyed into a distribution network through the media and use legitimization. Telecommunication instruments and new technologies have triggered off new art forms[2] that are becoming increasingly familiar and therefore desirable. For this reason, fruition has changed its characteristic of individuality and has taken on a semi-collective feature more appropriate to the user.

Technology has transformed fruition, not only by moving it from public to private, but also by making it an 'object'. Music, for instance, was once available only in concert halls and churches; nowadays it can be owned as an object (a record, a CD). Likewise, a tale, by going from oral to written, becomes a 'thing' (a book). In the case of the most advanced technology, the interaction occurs between man and machine, whereas the interaction between production and cultural organization takes shape through the mass media. There are several degrees of fruition, and therefore typologies of users, that can be classified as follows:

2 They are actually forms of social shaping [Qvortrup, 1998].

- Owners of cultural goods;
- Users of those goods in the place where they belong (churches or other public places, such as squares, train stations, etc.);
- Buyers of tickets in order to admire works of art, attend performances, and so on.

The user can be a consumer and owner both of art and goods, devoid of object substance, and of objects.

To exist as such, the user must 'know' and for this reason the information network promoted by the cultural organizations is pregnant and well structured. The tastes of the public take a back seat, but come to the fore when the economic results are concerned. The game of homogenization and differentiation is indeed originated by the hyper-segmentation of thematic issues and lifestyles. Such dynamics secures the creators of culture enormous control over their products. Over time, the 'cultural products' category has undergone an extension that has led users to choose niches of consumption. Paradoxically, users meet just because they share restricted and exclusive preferences[3]. The dynamics between supply of and demand for culture is such that success remains very uncertain [Griswold, 1997]. In fact, ultimate success mainly depends on the public, who has the last word. Complexity of consumption is sometimes unpredictable and this proves that the user is not only a knot in the net, but rather an individual that has to be convinced and consequently inserted in the complex consumption process. The user, then, has the power to influence production and for this reason innovation has to be predictable. The continuous input between communication, production and consumption causes circularity, promiscuity and changes that, by facilitating socialization, imply the acceptance of the product and of the organization that promotes it.

2 The work of art becomes a cultural commodity

The expansion of the tertiary sector through the proposition of cultural products, the need to organize cultural events in cities and towns, and

3 They meet also because they possess specific goods: hobbies, vintage cars, special cars, or other things.

thus the public bodies' duty to provide citizens with 'cultural packages' have forced the citizen/user to organize an *agenda setting* for works of art, exhibitions, biennials, and so on. Similarly to mass media information, which envisages that the person that knows what the others know should be informed, the most advertised, and therefore most attractive and 'in' events must be seen. This purely economic fact has an effect on the organization that must be able to remove that frame of 'artisticity', and so sacredness, that has surrounded works of art for centuries. Time, and the social dynamics that occurs in it, will decide what is worthy of being handed down to posterity.

The selection modalities of what is generically defined as *culture* consists in standards that cover the different strata of society and that will be mixed as happens in the so-called *cultural arena*, which is considered the arrival point of all the novelties that are conveyed out of their own setting. The cultural product made up of images, tales, and ideas pooled together from a wide range of sources becomes part of the *central sector*, as defined by Crane [1997], which highlights the separation of the product from its original setting and the preservation of a minimal part of the original information[4]. In a rigidly subdivided society, the *selector* can be assumed to expect from the user the *reproduction* of his/her own behavior, and that the same selection rules are valid for all the market understood as the meeting point of supply and demand. The term 'taste', whose acceptation was once shared by the assigner, the artist and the public, nowadays sees as part of its meaning the impulse to emancipation, the claim to one's own artistic preferences, the refusal to conform to pre-established models. In this sense, greater tolerance and helpfulness exist in our present society [Tessarolo, 2001; Verdi, 1998, 2003].

Each age has had different modalities to get the artist and the user to meet. The relationship between these particular communicators is complex, the complexity being due not to the encounter of two or more individuals, but to the encounter of two different 'visions of the world' [Zorino, 2003].

4 Those who create culture watch one another continuously in order to validate their aesthetic and political conceptions [Crane 1997, p. 26].

The public is attracted by things that are similar and so comprehensible, even if they are placed at different levels of symbolization. If not very long ago taste was shaped because it was possible to 'distinguish between a limited number of works', nowadays the word 'taste' has been replaced by the word 'consumption'. Such a change clearly indicates that current ideologies privilege consumer sectors whose preeminence is in no way established [Tessarolo, 1999].

Within each cultural sector there exist avant-gardes (which, in accordance with Crane, can be defined *resilient subcultures*) that attribute different meanings to culture. The avant-gardes strongly stress their own identity and, through differentiation, try to determine their social position. The products of the avant-gardes are selected by the media of the *central culture* to be spread in less provocative ways than those implemented by peripheral or urban channels. Large publics accept the selection of the media for their own consumption, which must not correspond to products that are too *new* since ambiguity would not lead to the anticipated outcome. Products must follow the *trend*, the taste of the time, and overcome the many stages necessary to get across the organizational system. The central media, which are strong selectors, tend to dominate the system as a whole and to control the normal proliferation of new, albeit secondary, cultural organizations.

The causes of the hyper-segmentation of issues and lifestyles in the public are not wholly understood. There is a tendency to the continuous differentiation of genres, although only the existence of new forms and cultural genres that are giving their creators some kind of control on their products has been determined. In any case, consumers represent the last knot of the cultural industry system, which aims to be defined in term of processes and practices and no longer in terms of things and objects. The reinstatement of the term 'user' is a sign of discontinuity, but serves also to find anew the traces of one's own history and to signal the birth of a previously missing cultural motivation.

A typical element of modernity is complexity, which is fragmented once it gets to the individual. People increasingly become individualized and develop into specialized niche users, all with the same restricted preferences. The distribution of culture – or, if you like, the request for a product – and its organization follow the wave of commercial success.

Griswold [1997] observes that, all the strategies devised by the major cultural industries notwithstanding, the uncertainty of success of a cultural product remains high. For instance, not all published books and songs are successful: the ultimate success of a cultural product rests with the consumers, the users, in other words the public. This is because, even if its authors can suggest the initial meaning of the cultural object, the public has the last word. In this, the public shows it has several degrees of freedom, which enlarge either according to the *cultural capital* [Bourdieu 1986] of the single individuals or to their expectations. The link between individual and society via the created symbols is so strong that 'the idea that aesthetic judgments are only and exclusively subjective is a vague and indeterminate commonplace' [Dahlhaus, 1970].

Within each sector, the existence of contamination and hybridization dynamics can be hypothesized to the extent that popular, learned and mass contents tend to mix and lose their precise features and give rise to new objects, which are problematic, as they are products for cultural consumption. The unpredictability of the demand, the surplus of creators, the inexpensive economy technology, the cultural industry are the transformation factors involved in the innovation of predictable commercial objects.

In order to create forms of identification between special groups of consumers and the objects produced, producers include in them a series of highly recognizable contents[5]. The ultimate users buy and actually utilize the goods and may allot them meanings that are different from those of their producers. The inputs that come form one direction and another produce circularity and thus continuous changes.

The search for new forms and structures, which arises out of the need for change and for identification, leads to the extreme specialization of the cultural processes.

It would then seem interesting to analyze the cultural multi-collocation of certain artifacts, which seems to derive from the reunification of several layers of observation that almost inevitably lead to the production of kitsch forms [Bertasio, 1998].

5 Let's think, for instance, of brand names, trademarks and the whole sponsoring system.

Besides the economic meaning, such products, which can be considered goods at all effects, have also a social and cultural significance. There is indeed congruence between supply and demand, but also the developing of new lifestyles and ways of consumption [Lunghi, 2003, p. 19]. Nowadays, every cultural event is marked by a combination of different genres that, if taken individually, can no longer configure self-sufficient universes. They are identities unconstrained by specific times, places, history and tradition [Giddens, 1994].

The novelty is that the arts tend to assert their autonomy within the mass market of show business rather than in the art market. Mass consumption confirms that the show is more important than art and that public consumption is more greatly appreciated than private consumption. In addition, information on the art events themselves has an artistic form.

3 Conclusions

The Weberian perspective that identifies the source of change within society is currently the most fruitful. It is unthinkable to try investigating the existence of some kind of cultural 'assistance' opposing the free market of culture. The many theories on culture [Crane, 1997; Griswold, 1997; Zolberg, 1994; Bourdieu, 1984; Ong, 1985; Negrotti, 1995; Bertasio, 1996; Verdi, 1996] offer very different possibilities of encounter between supply of and demand for culture. However, if culture within a service society is considered, art, like many other services, has detached itself from a physical location and also from a true human relationship. The move has been from object reality to extreme abstraction, where availability is continual and accessibility instantaneous. There is also the need for a differentiation that goes beyond the free market and the coexistence of new and old in all social domains [Semprini, 2003]. Current culture leads to a collective 'monologue' produced by the media. Such a monologue prepares and trains the individual to be transformed into a spectator, who, as such, is neither a participant in an experience nor an actor in an event, but a *user*, a consumer of some sort of artifact (an object, an image) in the formation of which he/she did not participate. All this poses the ever-present problem of 'bureaucratization', which is soul-destroying and seems to also involve symbolic needs, as it is oriented toward the differentiation of values and thus toward subjectivity. The doubt about soul-destroying or about livening up is always there.

With its participation in the work, the public establishes a social interaction that socially unites reception and production, so much so that it is sometimes impossible to distinguish the user from the artist.

It is fairly difficult to understand aesthetic social behavior because it is continually changing, in consideration also of the fact that new perspectives are, so to speak, forced on the public, even when it is convinced of the goodness of those that it already holds. The birth of mass production alongside traditional media leads to the dissemination of aesthetic convictions. However, even if medial aesthetics influences the way the world is observed and the way the individual interacts, socialization is the last step of acceptance of the new cultural forms technology produces [Qvortrup, 1998]. Individuals are not born users; it is society that will teach them to become 'full time' users.

If subjectivity is the principle of the modern age, from which the modern individuals draw their awareness of freedom, the multiplicity of viewpoints is valid also with reference to the multiplicity of culture, of the reference systems, and so on. Like subjectivity, multiplicity is a constituent part of the same culture, of the same reference system.

Bibliography

H. Becker, *Art Worlds*, University of California Press Berkeley, Ca 1982.

D. Bertasio, *Studi di sociologia dell'arte*, Angeli, Milano 1996.

D. Bertasio, 'Fra il bello e il brutto', in D. Bertasio (a cura di), *Immagini sociali dell'arte*, Laterza, Bari 1998, pp. 179-207.

D. Bertasio and G. Marchetti (a cura di), *Fra ombre e autoritratti*, Angeli, Milano 2000.

P. Bourdieu, 'Habitus, code and codification', in *Actes de la Recherche en Sciences Sociales*, 64, 1986, pp. 40-44.

D. Crane, *La produzione culturale*, Il Mulino, Bologna 1997.

C. Dahlhaus, *Fondamenti di storiografia della musica*, Discanto, Fiesole 1980.

A. Giddens, *Le conseguenze della modernità*, Il Mulino, Bologna 1994.
W. Griswold, *Sociologia della cultura*, Il Mulino, Bologna 1997.
S. Langer, *Sentimento e forma*, Feltrinelli, Milano 1965.
C. Lunghi, *Culture creole*, Angeli, Milano 2003.
E. Morin, 'Le vie della complessità', in G. Bocchi and M. Ceruti (a cura di), *La sfida della complessità*, Feltrinelli, Milano 1985, pp. 49-60.
M. Negrotti (a cura di), *Artificialia. La dimensione artificiale della natura umana*, Clueb, Bologna 1995.
W. Ong, *Oralità e scrittura*, Il Mulino. Bologna 1985.
J. Piaget, *Introduction à l'épistémologie génétique*, PUF, Paris 1950.
L. Qvortrup, 'L'arte e la società dei multimedia interattivi: lo shaping socio-estetico dei multimedia', in D. Bertasio (a cura di), *Immagini sociali dell'arte*, Dedalo, Bari 1998, pp. 145-177.
B. Sanguanini, 'Le arti tra opera, performance & informazione', in D. Bertasio (a cura di), *Immagini sociali dell'arte*, Dedalo, Bari 1998, pp. 89-126.
A. Semprini, *La società di flusso. Senso e identità nelle società contemporanee*, Angeli, Milano 2003.
M. Tessarolo, *La comunicazione tra dimensione umana e tecnologia*, Cleup, Padova 2001.
M. Tessarolo, 'Doppia utenza della musica leggera: la critica e il pubblico', in A. Porrello, *Modelli dell'utente nelle società avanzate*, IUAV, Università di Venezia 2003, pp. 177-188.
L. Verdi, *Habeas Corpus*, Angeli, Milano 1996.
L. Verdi, 'L'utente tra i luoghi e i non luoghi dell'arte', in A. Porrello, *Modelli dell'utente nelle società avanzate*, IUAV, Università di Venezia 2003, pp. 189-203.
L. Verdi, 'Il corpo nell'arte: forme e modelli di rappresentazione sociale', in D. Bertasio (a cura di), *Immagini sociali dell'arte*, Dedalo, Bari 1998, pp. 271-284.
R. Williams, *Sociologia della cultura*, Il Mulino, Bologna 1983.
V. L. Zolberg, *Sociologia dell'arte*, Il Mulino, Bologna 1994.
M. R. Zorino, 'Un modello di consumatore-produttore: la committenza nel Rinascimento italiano', in *Metis*, Vol. XI, 2003, pp. 1-34.
M. R. Zorino, 'L'utente da consumatore di servizi a cliente-committente', in A. Porrello, *Modelli dell'utente nelle società avanzate*, IUAV, Università di Venezia 2003, pp. 77-184.

Mariselda Tessarolo, Dipartimento di Psicologia Generale, Università di Padova

Laura Verdi

The places of the users

Abstract:

The perception of places, and therefore even of the specific art places, depends on the social representation (knowledge model) of individuals.

The perception of art (and its fruition) traditionally occurs through place perception. Paraphrasing Goffman, art places are 'those that position a "frame" around a stream of immediate events and determine the type of sense which each thing within that frame is to have'. However, which are the frame-models that define an art place and give it its identity? And which are their conditions of validity and the criteria for their formulation?

Nevertheless, virtual art diffused in the so-called noplaces loses in the end its own sense of place, and paradoxically scatters its own traces in a plurality of places (albeit without ever taking place). Hence a totally new dialectics between places and noplaces has originated, marked by a continuous changing of the macro, and then micro, urban landscape.

Keywords:
virtual art, *noplaces*, sense of place, dialectics of places/*noplaces*, changing of urban landscapes

1 The contexts of art fruition. The difficult road of democratisation

A non-secondary task of contemporary culture, in the balance between local and global, has become that of 'exploring the way mankind lives the global *ecumene*' [Hannerz 2001, pp. 27-29]. One of the ways to do this – and not the least – is entrusted to art and the places (albeit less and less specifiable) where it occurs.

Before beginning a brief analysis of this issue, moving towards the contexts of art fruition, some sociological categories useful to the purpose have to be recalled.

Great help on this will come from the important socio-anthropological category of *habitat of meaning*, proposed by Bauman and re-elaborated by Hannerz. Closer to the dimension of agency than Schutz's *overlapping province of meaning*, the concept of *habitat of meaning* takes into consideration the 'heterogeneity of the actors and of the organisational settings' as well as of the likelihood 'to extend and contract' the habitats themselves, which can intersect even if they carry different meanings. Thus, within the global *ecumene*, it is possible for several individuals who are in some respects totally different from one another, to share different several *habitats of meaning*.

The noble effort of democratising culture (idealistically placing all users in the same *habitat of meaning*) made by the ideologists of the multiple, has only originated, as Baudrillard ironically underlines, a new formula of social integration through undifferentiated consumption. Guaranteed by the media, it finds in the multiple the technical support that acts as an actual vehicle of social promotion.

The first who moved towards the democratisation of culture were the studies of the Frankfurt School in the past century.

Walter Benjamin (1966), first of all, observed that the different perception of art places is produced by the transition from the fixed image of the paintings to the mobile image of the movies (p. 43). The mobile image is not enjoyed through some form of contemplation, but, rather, it is perceived as the bullet-image of dadaistic memory, which stops the mind from placing images within precise frames. This the beginning of the phase of 'perception in absentmindedness', by which 'the public is the examiner, but an *absentminded examiner*' (p. 46) of works that have for ever lost their *aura*.

Adorno and Horkheimer maintain that art obeys a logic of instrumental rationality, produced by the alienation of art itself from the social and political settings and from cultural values. Within this perspective, art appears as manoeuvred by the dominant power and functional to the logics of the cultural industry. It is therefore still far away from a really democratic enjoyment.

Parsons's dream, whose theory of social action expressed itself mainly in terms of the common integration of values, finds here its most boastful expression. Having separated since 1950 the theory of voluntary action from the consideration of meaning and the historicity of the contingent action, Parsons illustrates its functionalism in the most abstract way in the AGIL integrationist model, where consumption can take on the main function of integration [Hall, 1999, pp. 113-114]. However, integration is understood as a sign of social promotion rather than a way of learning or actual cultural education [Baudrillard, 1976, p. 147].

Current studies on social sciences and mass communication are more oriented towards the analysis of meaning construction in language, in culture, in symbolic interaction and in memory schemata [cf. De Fleur and Ball-Rokeach, 1995, pp. 289-290]. If it is true that in a context of mass culture 'production tends to replace creation' [Cuche, 2003, p. 93], it does not necessarily follow that mass culture must

> result in a world culture […]. Mankind will not stop producing cultural difference. If a globalisation exists, it is that of the cultural 'goods' market, but […] recent studies show that it does not lead to homogeneity of consumption [Warnier 1999, in Cuche, 2003, p. 94].

In the face of a generally undifferentiated production of artistic goods, the only consolation is the continuing possibility of a differentiated fruition that can at last be considered, perhaps just for this, more democratic.

2 Places and noplaces. Art in the city

Art, like any other form of communication turned virtual, must contend with a changed sense of place. And, since space is socially constructed [Berger & Luckmann, 1969], the perception of places, and therefore even of the specific art places, depends on the social representation (knowledge model) of individuals.

The perception of art (and its fruition) traditionally occurs through place perception. Paraphrasing Goffman (1979), art places are 'those that place a "frame" around a stream of immediate events and determine the type of sense which each thing within that frame is to have' (p. 16). However, which are the frame-models that define an art place and give it its identity? And which are their conditions of validity and the criteria for their formulation?

Traditionally interpreted as fitting to carry incontrovertible cultural signs, art places are the Weberian webs on which the signs of art get entangled. The spider that has built them is the artist, who has also inscribed them at the same time into a 'context in which these events can be described in an intelligible way' [Geertz, 1987]. Nevertheless, virtual art diffused in the so-called *noplaces* in the end loses its own sense of place, and paradoxically scatters its own traces in a plurality of places (albeit without ever taking place). Hence a totally new dialectics between places and *noplaces* has originated, marked by a continuous changing of the macro, and then micro, urban landscape.

Meyrowitz's analysis, for instance, highlights how the relationship between the social experience (of the art user) and the physical place (its social construction and symbolic organisation) produces different group identities and different stages of socialisation and levels of authority [Meyrowitz, 1995, p. 13]. The electronic media cause the split between physical place and social place, and propose the user/place relationship in an apparently paradoxical way: that is to say, in the absence of the place [*ibid.*, p. 521]. Thus, Meyrowitz inscribes in a virtual and non-physical place (deterritorialisation) communications that happen without 'taking place'

The new, or poor, sense of place, triggered by the difficulty of maintaining (as in the egalitarian societies of hunters and collectors) many 'separate places' (also between public and private) tends to involve every individual in other people's affairs. Understood as nonfaithfulness to the territory it originates nomadism. The exchange between public and private causes the ensuing loss and change of the sense of place and its social representation. The extraordinary becomes common thanks to the spreading of cultural models without barriers.

It is then necessary to analyse how and why roles (e.g., the user's role) and places devoted to art [cf. Meyrowitz, 1995, p. 515] have changed. This happens in settings in which information is less linked to a physical place than to the *media*.

From distinct and segregate places which produced distinct and segregate knowledge, from the difficult access reserved to some social roles (*overlapping provinces of meaning*) there is a transition to a new social order in which undifferentiated places see the coexistence of opposite feelings and values (habitat of meaning), while culture becomes more and more *placeless*. The backstage leaps onto the stage and creates the need for new representational styles to come to the surface.

Architecture and the arts (even the applied arts) are full of examples of this: turning traditional places into *noplaces* and *noplaces* into places. The space of a city, in particular, if analysed in its social aspects and especially in those that refer to the action of individuals that enjoy art and make up the urban population, will provide suggestions for a novel reflection. By taking into consideration the relationships among users and between them and external subjects, and between them and social groups, movements, institutions and organisations, the analysis of the enjoyment of places and *noplaces* of art will excite a number of surprises.

Augé's terminology actually finds its more than semantic overturning where the changes we see almost unconsciously are functional and symbolic substantial modifications: train stations and European (for example London with its Tate Modern, or Paris with the Gare de Lyon, but Lisbon, too) and American undergrounds that are requalified as art containers. The same can be said for Naples with the projects for the 2002 Underground as an artistic station realised by Alessandro Mendini, with the collaboration of Kossuth, Kounellis, Cucchi, de Maria, Pistoletto, Paladino and many other young Neapolitan artists. Milan is an example, too, with the meeting between literature and means of transport achieved through the distribution of books collecting young writers' short stories, which can be read over the time of a few stops. Both these experiences have been able to transform means of transport and mobility places (already *noplaces*) into art galleries and meeting points.

I would also like to mention the outstanding exhibition '*Shopping: One hundred years of art and consumption*' held at the Schirn

Kunsthalle in Frankfurt and closed on 1st December 2002, where the central point was a re-reading of the relationship between art and consumption in the Nineteenth century. The exhibition, organised by Max Hollein and Cristoph Grunenberg, collected the works of 60 modern and contemporary artists from Warhol to Christo, from Richter to Duchamp. All that was accompanied by a selection of films on the topic of shopping. Among the works exhibited there was a real supermarket with real products. And I would also like to mention, again in Frankfurt, at the airport, Terminal 1, departure lounge 2, the exhibition '1 Euro market', where in a 500 square metres area 4,000 plastic products were for sale for the price of 1 Euro each. How not to perceive the provocative echo of the theories propounded by the School of that same town, which more than half a century before had already caught the contradictions existing in aesthetics as well as the risks of reification?

Similar projects are the 'designer's stations' for the strengthening of high speed all over Italy: and so Bologna Central, Roma Tiburtina, Turin Porta Susa and Florence Santa Maria Novella, the contest winners, propose themselves as new monuments to contemporariness.

Last, but not least, I would like to mention here the example in the making of the *Bienal de Valencia* which was devoted last year 2003 to the Ideal City. Organised by Lorand Hegyi, it gathers the ideas of tens of artists from all over the world on how to rebuild through art some buildings partially abandoned and demolished by the war. Ninety *solares* were entrusted to 40 artists to attempt the reconstruction of the link between the place one lives in and the 'big house' which the city is. The projects are of great artistic, as well as sociologic, anthropologic, poetic and archaeological interest. Thanks to this exhibition it is possible to see contemporary art outside the traditional museum settings, in fact anywhere and everywhere; this helps us to open our mind to new ways of art fruition. Hegyi's words 'Surreal and unusual images are created in which everyday reality suddenly transforms itself into something very different, till then unknown' remind us of Bachelard (1999). 'It is a matter of living the unlived and opening to a language opening' (p. 20) as 'there is no poetry without absolute creation' (p. 21).

In contrast, places traditionally connoted (art galleries, but also restaurants and shops) present themselves as *time places*, symbols of the celebration of the ephemeral.

Certainties, be they economic or cultural, have gone and the western tendency to the *openclose* mirrors the loss of concreteness of many traditional places, whose sites are positioned in spaces destined to a brief duration. Here the works of art on show are themselves destined to crumble, as is highlighted by the Paris sociologist and trend scholar Vincent Gregoire. There are elusive photo and painting exhibitions whose beginning and end are unknown in an ever increasing dramatisation of accelerated times, in which novelty and decay excitedly follow one another.

Architects, artists and designers continually propose settings that seem planned to excite emotion rather than last. Perhaps this is the ridge on which today's distinction between design and contemporary art is measured: the former gratuitous and destined to a brief duration, the latter expensive, often difficult to understand and communicate, but destined to last.

And yet, museums still stand fast. Once places of art consecration, they recycle themselves – so to speak – for the public at large as gigantic containers without *aura*. Having broken down the barriers between *elites* and masses, competent and curious, museums have created bazaar-spaces, shopping centres fit for buying and consuming (forbidden as regards works of art, but reintroduced as regards their multiples). Thus discarding – at least apparently – the distance between privileged and unprepared audience, they are places desecrated by the democratisation of the architectural forms (less rigid and often more inclined towards the curve and the least expected boldness, even in the use of materials, as in Frank Gehry's 1997 Guggenheim museum in Bilbao).

These spaces, nomo- and mono-logically appointed to contain forms of socially isolated and less communicative beauty, had been conceived in the Nineteenth century with a spirit tending to a storage classification and aesthetic enjoyment which was never separated from labour: almost as if to remind the audience that beauty is compelling and, if excessive, it must somehow have something to be forgiven for, the puritan mentality that creates the museum allows very few spaces for reflection and pause. Only a few uncomfortable seats interrupted the breathless visit to the traditional museum: the building was austere, massive and meant to discourage outsiders with its outward appearance. Hence the British Museum, the Louvre, the Uffizi, the Kunsthistorisches, the Prado, etc.

The prototype of the contemporary museum, changed in its architectural and functional conception, uses light, materials, sounds and colours in an evocative way: by socially constructing differently functioning spaces in every place, and opening continuous gaps to incorporate the *noplace* into the place.

3 Places beyond places. Liquid beauty?

Mediated by the *new media*, the social representation of art places becomes fluid: art fruition (unlike reading which requires solitary, rather than socialised use) can be transformed by the auxiliary help of the media (screens, videos, etc.) within the places devoted to art and become a new form of sharing.

It is also in this way that the stimulus to create newer and newer needs, which Simmel had already recognised as a characteristics of the urban setting, finds a brand new modality of satisfaction [Simmel, 1995, p. 52]. Now more than ever, thanks to the unstoppable expansion of out-of-place places, 'a town exists only in the set of effects that go beyond their immediacy' [*ibid.*, p. 51]. Such effects, then, go well beyond what Simmel, with his extraordinarily sharp analysis, could ever have imagined a century ago.

Among these effects, what changes its valence to become liquid [Bauman, 2002] and diffuse, albeit not ubiquitous, is the idea itself of beauty. A beauty that, stimulated as liquids are by external pressures, cannot be long-lasting. Not even when it relates to places. Like nation-states and individuals, who are always pressed for new competencies and values and therefore cannot maintain their identity, beauty finds itself 'forced' into a new frame, which takes away its characteristics of universality and eternity, no longer using supreme taste as its reference point. Ever modifiable, beauty can be traced not to objects but to events. Unlike Bauman, Susan Sontag maintains that 'the fact that beauty is no longer a criterion to judge art does not point to a decline of its authority. It is rather an index of the declining conviction that something called art still exists'. Sontag, on the contrary, believes it right to talk about the beauty of art that gets back its own solidity 'as the necessary judgement to give sense to the greater part of our energies, affinities and admirations'. Beauty cannot be detached from ethics, as Kierkegaard and Tolstoj thought: 'the aesthetic projects is almost in itself a moral project'. And then, the beauty of art cannot always be only consolatory. As Hegel said, it is better than the beauty of nature because it comes from the spirit.

Beauty becomes liquid also slipping out of natural places to materialise in other social representations: in other artificial places or, again, in *noplaces*, fake natural paradises, like parks with rare or exotic plants or animals. Beauty becomes liquid also by shifting onto implausible models of artificially mutant bodies, always hovering between nature and culture, being individuals within places (*Sein*) or creatures in the making within *noplaces* (*Dasein*).

4 Landscape, city and art

Reached and identified with the uncertain compass of contemporary beauty, places and *noplaces* are the just as uncertain frame for art in the city. Still acknowledging an important relationship between city and landscape that can be organised through instruments of planning (which requires consensus), the question arises of how it is still possible to pursue consensus for the planning of the landscape. Perceived not only as a productive space, but also as a symbolic good, a collective cultural reference point, a sign of shared intents, free time and relax, the traditional landscape is in sum a 'paysage patrimoine'[1].

1 From the analysis of about 9,000 photographs sent in 1992 to a contest entitled 'Le paysage préféré des Français [The French people's favourite landscape]', Françoise Dubost, who teaches sociology of art, evinced that the favourite landscape of the French was mainly rural and agro-pastoral.

And what about the less traditional city landscape? The relationship between landscape, city and art can be defined also through the use that is made of space and art in public places by *inhabitants, commuters*, city *users* and *metropolitan businessmen* (categories all used by Martinotti). Art fruition, arranged in different places and at different times for each of these typologies of inhabitants, has stopped organising itself within recurrent and recognisable frames: outside the places of the city reserved only to its manifestations, art overflows from places and involves the surrounding landscape as well as places unthought of for its manifestation.

An example I can think of is the important Expo 02: a colossal fair held between May and October 2002 in Switzerland that involved two linguistic regions, three lakes and five cities, with artistic and engineering wonders (worth 900 million Euros) that were to disappear at the end of the Expo.

Visionary installations joined the earth and the lake in Neufchatel; sharp steel towers disrupted the hills of Biel, by catching and reproducing the far away sounds of the lake; the Cloud of the Yverdon-les-Bains *Arteplage*, a 100 metres platform, released 33,000 little magical jets of water; marvels that had taken seven years to be ready, created by great architects only to disappear six months later, swinging in the unsteadiness between *noplaces* and places.

5 Living responsibly

Talking about art in the city and about places and *noplaces* for its manifestations, I would like to add something more about the meaning of experiencing a place together and share its cultural use through its space.

An important preliminary distinction is to be ascribed to the philosopher Curi (2000), who differentiates in the moment of the city founding the expression of the *sunoikìa* (the most elementary dimension of living together) from that of the *pòlis* (the actual city) through the *politiché téchne*, as a sharing of responsibility as well as of privileges.

A distinction similar to Galimberti's (2002), who, making reference to the ancient classic priority of a political acting (*pràxis*) over the technical doing (*pòiesis*), must then bitterly acknowledge the overturning of the very same relationship in the century just over. In it, the technical-economic doing and the functional aspect of doing prevail over the symbolic aspect of the political acting, reducing our cultural horizon (becoming more and more a-topical) to a desert of goals. The city in which building prevails over living [Gregotti, 1996] breaks its connections with the specific identity and loses its territoriality, while the idea itself of the place as representation of a social order is challenged. And so the sense of the Frankfurt criticism to the *aura* ideology, eluded by the technical doing through the multiple, will have to be applied today less to the triumph of the symbol (politics) than to the pretence of technique.

Also touched is the crucial point of *cum-munitas* [Esposito], expression that we will not translate as *Gemeinschaft* without implicit reference to the sharing of responsibility. The latter is more typical of *inhabitants* than of *commuters*, *city users* and *metropolitan businessmen* who are used to a discontinuous relationship with the places of the city that puts them in a position of *im-munitas* rather than *cum-munitas*, by exonerating rather than involving them in a shared duty. These are 'immune' citizens, too, who in their *disembedding* [Giddens], bring up day after day the question of their cultural identity (*re-embedding*) and their relationship with the users of places (cities first of all) and the art scattered in them. We cannot ignore the desolating perspective awaiting us around the year 2030, when 60% of the planet's inhabitants will live in an urban dimension[2].

Places, cities and art, devoid of symbolic and political goals, run the risk of confining themselves into a self-referentiality without meaning and without future.

2 See data presented at the conference 'La nuova cultura delle città' [The new culture of cities], organised in Rome, 5-6-7 November 2002, at and by the National Academy of Lincei and the CNR (Italian acronym for National Research Centre).

Bibliography

A. Abruzzese and D. Borrelli, *L'industria culturale*, Carocci, Roma 2000.

T. Adorno, *Einleitung in die Musiksoziologie. Zwölf Theoretische Vorlesungen*, Suhrkamp Verlag, Frankfurt 1971, It. transl., *Introduzione alla sociologia della musica*, Einaudi, Torino 1962.

M. Augé, *Non-lieux*, Seuil, Paris 1992, It. transl., *Nonluoghi. Introduzione a una antropologia della surmodernità*, Eléuthera, Milano 1996.

G. Bachelard, *La poétique de l'espace*, Presses Universitaires de France, Paris 1957, It. transl., *La poetica dello spazio*, Dedalo, Bari 1999.

J. Baudrillard, *La societé de consommation. Ses mythes ses structures*, Gallimard, Paris 1974, It. transl., *La società dei consumi*, Il Mulino, Bologna 1976.

J. Baudrillard, *Le crime parfait*, Editions Galilée, Paris 1995, It. transl., *Il delitto perfetto*, Cortina, Milano 1996.

Z. Bauman, 'A Note on Mass Culture: On Infrastructure', in D. McQuail (ed.), *Sociology of Mass Communication*, Penguin, Harmondsworth 1972, pp. 61-74.

Z. Bauman, *Intimations of Postmodernity*, Routledge, London 1992.

Z. Bauman, *Liquid Modernity*, Blackwell Publishers, Oxford USA 2000, It. transl., *Modernità liquida*, Laterza, Roma-Bari 2002.

P. L. Berger and T. Luckmann, *The Social Construction of Reality: A Treatise in the Sociology of Knowledge*, Anchor Books, New York 1967, It. transl., *La realtà come costruzione sociale*, Il Mulino, Bologna 1969.

M. Berman, *All That is Solid Melts into Air: The Experience of Modernity*, Simon and Schuster, New York 1982, It. transl., *L'esperienza della modernità*, Il Mulino, Bologna 1985.

W. Benjamin, *Das Kunstwerk im Zeitalter seiner technischen Reproduzierbarkeit*, from W. Benjamin, *Schriften*, Suhrkamp Verlag, Frankfurt 1955, It. transl., *L'opera d'arte nell'epoca della sua riproducibilità tecnica*, Einaudi, Torino 1966.

W. Benjamin, *Das Passagen-Werk*, Suhrkamp Verlag, Frankfurt 1982, It. transl., *Parigi, capitale del XIX secolo. I passages di Parigi*, Einaudi, Torino 1986.

H. Blumer, 'An Appraisal of Thomas and Znaniecki's "The Polish Peasant in Europe and America" ', in *Social Science Research Council Bulletin*, 44, New York 1939.

V. Codeluppi, *Consumo e comunicazione. Merci, messaggi e pubblicità nelle società contemporanee*, Angeli, Milano 1989.

F. Crespi, *Manuale di Sociologia della cultura*, Laterza, Roma-Bari 1996.

D. Cuche, *La notion de culture dans les sciences sociales*, Editions de la Découverte, Paris 2001, It. transl., *La nozione di cultura nelle scienze sociali*, Il Mulino, Bologna 2003.

U. Curi, *Filosofia come guerra*, Bollati Boringhieri, Torino 2000.

M. L. De Fleur and S. J. Ball-Rokeach, *Theories of Mass Communication*, Longman, New York 1989, It. transl., *Teorie delle comunicazioni di massa*, Il Mulino, Bologna 1995.

P. Dickens, *Urban Sociology: Society, Locality and Human Nature*, Harvester Wheatsheaf, Hemel Hempstead 1990, It. transl., *Sociologia urbana,* Il Mulino, Bologna 1992.

R. Esposito, *Communitas. Origine e destino della comunità*, Einaudi, Torino 1998.

R. Esposito, *Immunitas. Protezione e negazione della vita*, Einaudi, Torino 2002.

G. P. Fabris, *Consumatore & mercato. Le nuove regole*, Sperling & Kupfer, Milano 1995.

U. Galimberti, 'Il futuro si chiama non luogo', *La Repubblica*, 18 July 2002, pp. 42-43.

A. Gasparini, J. R. Logan and V. Mansurov (eds.), *Riqualificazione e hinterland delle grandi città*, Angeli, Milano 1994.

A. Gasparini, *Luoghi, città, società*, Carocci, Roma 1999.

C. Geertz, *The Interpretation of Cultures*, Basic Books, New York 1973, It. transl., *Interpretazione di culture*, Il Mulino, Bologna 1987.

A. Giddens, *The Consequences of Modernity*, Polity Press, Cambridge 1990, It. transl., *Le conseguenze della modernità*, Il Mulino, Bologna 1994.

E. Goffman, *Encounters: Two studies in the Sociology of Interaction*, Bobbs-Merrill, Indianapolis 1961, It. transl., *Espressione e identità*, Mondadori, Milano 1979.

V. Gregotti, G. Marzari and L. Figini (eds.), *Opera Completa*, Electa, Milano 1996.

J. R. Hall, *Cultures of Inquiry: From Epistemology to Discourse in Sociohistorical Research*, CUP, Cambridge 1999.

U. Hannerz, *Exploring the Inner City*, Columbia University Press, 1983, It. transl., *Esplorare la città*, Il Mulino, Bologna 1992.

U. Hannerz, *Transnational Connections: Culture, People, Places*, Routledge, London & New York 1996, It. transl., *La diversità culturale,* Il Mulino, Bologna 2001.

F. La Cecla, *Perdersi. L'uomo senza ambiente*, Laterza, Roma-Bari 1996.

F. La Cecla, *Non è cosa. Vita affettiva degli oggetti*, Eléuthera, Milano 1998.

T. Maldonado, *Critica della ragione informatica*, Feltrinelli, Milano 1998.

F. Martinelli (ed.), *Città e campagna. La sociologia urbana e rurale*, Liguori, Napoli 1981.

F. Martinelli, *Le società urbane*, Angeli, Milano 1990.

G. Martinotti, *Metropoli*, Il Mulino, Bologna 1993.

D. McQuail, *Mass Communication Theory: An Introduction*, Sage Publications, London 1983, It. transl., *Sociologia dei media*, Il Mulino, Bologna 1986.

J. Meyrowitz, *No Sense of Place: The Impact of Electronic Media on Social Behavior*, Oxford University Press, New York 1985, It. transl., *Oltre il senso del luogo*, Baskerville, Bologna 1995.

G. A. Micheli, *Dentro la città. Forme dell'habitat e pratiche sociali*, Angeli, Milano 2002.

R. Park, 'The city: Suggestions for the Investigation of Human Behavior in the Urban Environment', in *The American Journal of Sociology*, XX, March 1915, pp. 577-613, It. transl. in R. E. Park, E. W. Burgess and R. D. MacKenzie (eds.), *La città*, Comunità, Milano 1999.

T. Parsons, *The Social System*, Routledge and Paul, London 1951, It. transl., *Il sistema sociale,* Comunità, Milano 1965.
P. Rossi, *Paragone degli ingegni moderni e postmoderni*, Il Mulino, Bologna 1989.
M. Scheler, *Die Formen des Wissen und die Bildung* (1925), in M. Scheler (ed.), *Philosophische Weltanschauung*, Cohen, Bonn 1929.
H. Schwartz and J. Jacobs, *Qualitative Sociology: A Method to the Madness*, The Free Press, New York 1979, It. transl., *Sociologia qualitativa*, Il Mulino, Bologna 1987.
G. Simmel, *Die Großstädte und das Geistesleben* (1903), from M. Landmann and M. Susman (eds.), *Brücke und Tür*, Koehler Verlag, Stuttgart 1957, It. transl., *La metropoli e la vita dello spirito,* Armando, Roma 1995.
S. Sontag, 'Cosa dico quando dico "bellezza" ', *La Repubblica*, 13 July 2002, pp. 36-37.
J. P. Warnier, *La mondialisation de la culture*, La Découverte, Paris 1999.
V. Zolberg, *Constructing a Sociology of the Arts*, CUP, Cambridge 1990, It. transl., *Sociologia dell'arte*, Il Mulino, Bologna 1994.

Laura Verdi, Dipartimento di Sociologia, Università di Padova

Maria Romana Zorino

The client-assigner of cultural artifacts

Man is placed in time to have time and not to get somewhere with his legs before he gets there with his heart.

Karl Kraus, *The last days of mankind*

Abstract:

Nowadays, from the technological and economic point of view, in many product sectors it is possible to produce ad personam, *which allows the client to became the assigner. In order to develop an ideal type of assignement relationship, I have interviewed the artisans from Naples that produce high-class clothing for the elitist consumption. It appears that the made-to-measure object is part of a symbolic world that integrates person, society and object. The three elements become one thanks to the alchemy of a ritual process, of which the tailor is the priest. The attempt, computer based, of mass customisation to reproduce this model is vain. The mass customization offers to the client several variants of modular assemblages that can be chosen on a virtual catalogue, following the model of the web page. The proposed differentiation is not a social, but a technical relationship. A model that is based on holistic unity cannot be 'copied' in terms of modular fragmentation. Neither is it possible for a purely technical-functional relationship to assume a social and/or ritual function. It would be far more interesting to propose a new pattern of assignment computer based, which could stimulate and enable clients to exercise their own creativity; however, this is hardly done in the ideal type defined in this paper.*

Keywords
assignement, Neapolitan tailoring, mass customization

1 Entrepreneur-client *vs.* selector-client

The introduction of computers in industrial production was one of the causes contributing to the 'democratization' of industrial capitalism and allowing the transformation of the production-consumption circuit. The 'democratization' consist in:

- The spreading of entrepreneurship following the fall of the barriers that hindered entrance in the entrepreneurial world because of the costs of the Fordistic production equipment. Part of this new entrepreneurship are a) the myriad of small entrepreneurs belonging in some way to dependence chains (the 'filiera' of the Italian industrial districts; the sub-contracting pyramid of the Toyota System; the Porterian value chain of globalization); b) the many formally 'self-employed' workers; c) employees with ample margins of autonomy. Creativity and flexible management of own resources are the main features of this entrepreneurship, which doesn't guide capitalistic development strategies (to establish the crucial centers of capitalistic power we should look to the financial capital).
- The change in the nature of firm. The possibility of real time communications and the need for flexible relationships with the environment instead of the expansive, rigid (and vulnerable) planned relationships have led to enterprise decentralization and to greater interpenetration with the environment. The boundary with the environment has become weaker and it is difficult to say what is part of the enterprise and what is not. Theories such as those of the network-enterprise (Vaccà, 1989) and of the *milieu innovateur* (Camagni, 1989) express this new nature.

Within this context, also the consumer's place is potentially changing into that of a client-entrepreneur that becomes part of the firm.

Nowadays, from the technological and economic point of view, in many product sectors it is possible to produce almost *ad personam*, which allows the client to become the assigner. A strategy that would be particularly advantageous for businesses, because it would allow to 'predict' demand and gain the trust of the client, both of which avoid the disavantages of planning. On the other hand, when a client becomes an assigner he/she will also participate in determining the course taken by the firm both as regards the static (management and organization of resources) and the dynamic (innovation) allocation. That is not enough. The assignment relationship profoundly changes the management itself of the capitalistic market because it reverses the driving function of supply in the driving function of demand.

Assignment questions also the nature of the consumer. He/she could a) access elitist ways of consumption no longer as the Veblenian imitation of the material symbols of a process, but as the acquisition of a style of life; b) come out of the Baudrillardian passiveness of the consumer of signs that impose their own grammar.

Actually, these were the promises of *mass customization*. However, *mass customization* does not target a client-assigner, but rather a client who is offered several variants of modular assemblages that can be chosen on a virtual catalogue, following the model of the web page, with the assistance of a human interface. It is a pseudo-differentiation within an asocial inclusion offered to a solitary crowd. The proposed differentiation is not a social, but a technical relationship and it contributes to the further fragmentation of individuals. It could have a sense if it allowed people to express their own creativity. However, the client cannot deviate from the patterns of the production software, and the producer's innovation is itself stifled. The countless possibilities of variations on the theme hide the fact that the door (which had always been left potentially open by the artisan and the industrial production alike) is closed to the radical variation of the theme. However, years of mass consumption have not passed in vain. From recent focus groups held with a number of final-year undergraduate students it has emerged that most consumers are not at all able to propose significantly new products.

2 The ideal type of the assignment relationship: the Neapolitan elite tailoring

What is the assignment relationship? In the elitist consumption it has never disappeared and from it I will try to develop an ideal Weberian type. To this aim I have interviewed some of the few artisans that produce high-class clothing for a very restricted number of lucky people. The interviewees are from Naples, where there is a high concentration of both clients' and producers' competencies (according to Assomoda,

Milan is the capital of the prêt-a-porter, Rome the capital of designers' clothes and Naples the capital of the 'made-to-measure'). Among the interviewees there are some very recent rising stars (Dell'Ero and Marolla); other less recent ones (Panico, Solito, Martuozzo); well-established, long-standing names (Rubinacci, Marinella); glories of the past (Attolini); timeless artists (Talarico). Alongside these 'production artisans', who are also service artisans, I have placed a pure 'service artisan' (Rossi).

Among their clients we can find politicians, professionals, entrepreneurs, a few actors and some remains of nobility. Their age goes from 35 to 70 with a peak at around 50. The clientele is both national and international, but the core is made up of Neapolitans, who are demanding experts.

The client and the producer pivot around a center of interest: the wellbeing of the client. The centrality of the client is highlighted by the respectful use of the term 'gentlemen' and the worry about their 'pleasure', 'enjoyment', 'wellbeing', 'comfort', 'ease', 'feeling fulfilled'. Anna Martuozzo, while talking about a collar that might not suit a client, worries about the client's possible 'suffering'. Within this framework both parties act in order to enable the clients to espress their needs.

Clients express their desires in a number of verbal and non-verbal ways.

A The simple choice of going, say, to Solito's or Panico's expresses preference for an up-to-date version of the Neapolitan jacket. Those who choose Claudio Attolini's atelier think of the old Neapolitan jacket worn by De Sica or Eduardo de Filippo (Vincenzo's clients, Claudio's father, whose style has been followed by the son). Should these clients be proposed models different from *THEIR* jacket, they would feel offended. Rubinacci, instead, as we read in an interview he recently gave the magazine 'Monsieur', believes in the cooperation with a creative client.

B Particularly elegant clients or those with a marked personality do not necessarily arrive with an idea or a specific project. They challenge the artisan to realize *THE* product that expresses them and that is up to the tailor's fame. This requires the tailor to create an *ad hoc* product, which is not the Neapolitan jacket in the version for which he is famous (up-to-date, internationalized, or conservative), but rather that very client's jacket (which can become a further trademark for the tailor). It is the case of the challenge posed to Panico by a nonchalantly elegant client, who got this tailor to produce a jacket cut only on the shoulders.

C Some requests go beyond the more diffused patterns, like those made to Dell'Ero and Marolla, and can involve either the whole structure of the product (a shirt unique as regards shape, material and color combination), or only some out-of-the-ordinary details (like the umbrellas with colored lights that Talarico was asked for).

D There are simple requests for particular details that make, say, a shirt special only to the wearer, like for instance a string to keep it straight, as was modern some time ago.

E There are preferences of style: a smooth shirt with a light material or the opposite.

F Lastly, there are a thousand variants to the finishing touches: the type of buttonholes, their positioning; internal or external initials, on the cuffs or on the collar points, and so on.

However, there is a boundary that none of these artisans of clothing is prepared to cross. A boundary of good taste and bon ton that nobody goes over: 1) because the client must not be open to ridicule and 2) because the tailor's image would suffer. These tailors 'allow' extravagant originalities only to some particularly elegant clients. Panico is drastic: things that would suit that particular client would be 'a shame' on others. The clients that are able to manage originality with grit and determination become reference models (Dell'Ero and Marolla discuss the daring but successful variants elaborated by particular clients). They 'allow' to clients with an outstanding personality the excess that on others would be a flaw. Rubinacci points to Churchill, who, rather than hiding his already imposing size, highlighted it by wearing checked clothes.

And here we find the other component of the relationship, the artisan's function as a guide and counselor:

A There are the 'neophytes' of the made-to-measure, who are taught how to dress by the tailors (from the accessories to the barber's address), until gradually the person also changes his posture (Solito). Some clients seem to ask for their 'features to be changed' (Panico).

B However, most clients already have a 'culture' of the made-to-measure and/or are elegant. The tailor orients them in the novelties, the choice of fabric and matching patterns contextualized to the individual, roles, social situations, events, functionality, personal wellbeing, the social function, and so on. And this he does in several ways, by letting the client 'see'

> some alternatives and 'touch' the material, by pre-selecting the range of possible choices in accordance with his knowledge of the client.

To be able to 'play midwife' to the client's wishes, the tailor begins by chitchatting: this allows him to learn about his client's character, his moving-about habits, his job, his tastes, the event for which the clothes will be needed. This survey will enable him, if he is a service artisan (as Rossi), to pre-select the fabric to propose to the client or, if he is a production artisan, to define the purpose of the clothes, the stress they will undergo, and so on. The artisan (like an anthropologist) tries – via subsequent approximations – to get into the client's world. This exchange is partly verbal and partly non-verbal. The material symbol of this immaterial pathway is the rite of trying the garment on, where the interpretational errors can be redressed.

The outcome is a garment that is a) integral part of an individual perceived in a holistic way; b) part of self-care. For his self-care, the client relies on the artisan in the same way as he would on a luxury 'beauty farm' (I am borrowing this term from Rubinacci). This idea of taking care of oneself is shared by all the interviewees, from Martuozzo, who worries about the client's wellbeing, to Panico, who says it is his prestige to create clothes that make his clients comfortable. The idea of care is expressed in many ways: it is related to physical wellbeing, psychical wellbeing, comfort, nonchalance, social prestige. The Neapolitan jacket itself has characteristics of comfort. Attolini shows that the position of the trousers pocket proposed by his father allows the hand to slip into the pocket 'naturally'. For its lightness, Solito's outfit is known as 'Breath of wind'. Panico notices that a ready-made garment is an armour, whereas a made-to-measure one integrates with the body. The ideal Neapolitan jacket is without much wadding; it must hang perfectly and be a second skin at the same time.

The atelier is also a place where people go to relax, to unload, to be cared for. There, you can sip a coffee and have a chat. Antonio Rossi says that his client is like a guest in his house; Rubinacci mentions a beauty farm. Some tailors say that their clients go to them to get rid of their stress, but also to give vent to their manias.

Wellbeing is referred to an individual perceived as a whole, where the soma, the psyche and the environment integrate. To be really good, the garment must become part of this unity. 'It should be like a good marriage', says Panico talking about body and garment understood as psyche and soma, and then, to express the wholeness of a person, he talks of 'dressing the soul'. The definition that everybody gives of the made-to-measure garment as 'comfortable' and the insistence on the naturalness necessary to wear it converge toward a connotation of 'comfort' that joins the sense of wellbeing with the unitary integration of body, personality, way of moving, life environment.

All that is achieved through expert work backed by many years of training. The result of naturalness in a garment is somewhat like the 'sprezzatura' (sophisticated spontaneity) of the courtier of Baldassar Castiglione, which has a lot of work behind. And the client plays his part too, of course. The interviewees deny that elegance can come from clothes. Choosing an outfit is a question of taste, not of elegance. Elegance is something that the individual has (or does not have) and it is mostly the result of a culture and of a habit.

Both parties cooperate in building the reference outline within which to move. The tailor is ahead in the knowledge of innovations about fabrics and fashions and provides the user with a compass, so to speak, to enable him to move in this setting, thus simplifying a complex system. However, with his acceptances or refusals, it is the user who allows the tailor to continually set the direction of that compass. It would also be an error to think that the bond between tailor and client is a simple relationship between two people. The made-to-measure clients are part of a social group, or at least wish to be and the made-to-measure clothes are part of that social inclusion. The individuals that patronize that tailor communicate between one another also by using that tailor and, in turn, the tailor gets a continuous input that enables him to build and de-build the diktats of elegance together with is clients. It is also through the tailor that clients can create their Simmelian course of inclusion and distinction as regards their reference group and the society of which they are members.

Integral part of the person-object osmosis is the dignity of the object itself; not any object, but an object with its own characteristics. To this contribute quality of the fabric, care for details, adjustment to the individual, but also far less material elements, such as knowing the 'life' of the object, from its origin to the phases of construction, to, last but not least, the waiting time. Unlike industrial objects, objects of refined artisanship give off an aura that makes them unique, important and cared for. They are not mere vehicles of desires, always evoked and never satisfied. They actually realize the desire and are a source of pleasure. The pleasure of choosing (Marinella says 'we don't sell a tie, but en emotion'), the pleasure of participating in creating the object (Dell'Ero talks of clients that 'amused themselves in …'), the pleasure of contemplative enjoyment (a client of these ateliers actually mentioned the physical pleasure of looking at the accuracy of the finishing touches) It is a hedonistic feature different from the advertising images that excite emotions and desires that will never be satisfied. It is not the negative pleasure of consumerism, which plays on the continuous frustration of an eternally postponed satisfaction. It is a positive pleasure that becomes part of the object structure and is recalled and renewed by it. A pleasure shared by client and tailor alike. The artisans I have interviewed love their work, are satisfied with what they produce. They talk about it with love, with pride, … with pleasure. Talarico looks upon his umbrellas as the children he doesn't have. He kisses them before sending them around the world.

Bauman (2004) maintains that the product becomes less important when the waiting is taken away from the wish. In the 'made-to-measure' the waiting time exists. It is time to metabolize, time devoted to oneself and to the object. For all these reasons these artisan artifacts last in time, they are indeed timeless. Looked at, cuddled, appreciated with pleasure by their owners, used in time, perhaps always the same ones (Marinella says that the ties that are actually worn are very few compared to the thousands his clients have in their wardrobes; the artisan-tailors themselves, as elegant as they are, wear 'old' clothes) they are object of care, attention and restoration (Attolini alters clothes if his client has 'changed' size; Marinella removes stains from ties; Talarico mends broken umbrellas). One example for all is the ceremony of the 'cirage' (*polishing*) of one's favorite shoes, which is quite the thing in the elegant clubs like 'The knights of the seven doors'. They are objects made to last, even if not destined to be worn much, like the formal clothes, and open to the client's total look; a client that 'knows' them even in their hidden places, like Capucci's elaborate linings and Talarico's umbrella ferrules. Those who 'own' these objects are indeed rich, because they do possess something, unlike the 'poor' who live in the realm of signs and wishes disjointed from things.

Also this 'important' object that makes you rich is a holistic structure that joins together raw material, form and destination. And it is not only a question of choosing fabrics suitable to forms ('you must not *abuse* a fabric' to make it what it isn't, says a wedding-dress maker), but of creating a unique object comparable to a work of art. A product that is harmonious in itself, not simply functional to something else. This unitary philosophy is visible in the aesthetics of the garment, whose details make up an integrated whole that could not be realized through a modular procedure. To create comfortable trousers, Claudio Attolini has not only modified the position of the pockets, but has also reinforced the front, so that comfort would not also mean deformation. Similarly, the dart making a line in the jacket to the bottom of it ends up perfectly in the boat-like line of the pocket. Panico's seamless jacket plays on a cut that considers the object as a single whole.

The harmonious connection between the three elements – person, object and environment – is the result of a hermeneutic reading and of the expert construction of an ecological adjustment. The ecological adjustment is creative even when it invents nothing, because every time it is a situated action that aims to the harmony of the whole. The situated action itself is an action as a whole, not translatable into modular parts. The unit that is formed from time to time is not a set of parts, but a co-growth of units. This alone would be enough to establish the impossibility of transitioning from the model of elitist artisan assignment to mass customization.

In addition, person, object and environment become one thanks to the alchemy of a ritual process, of which the tailor is the priest. Like all rites, it includes/excludes, creates the area of sacred/profane, attributes exceptionality to individuals and ritual things. The times of the made-to-measure, the gestures (many clients always want to try on their clothes even if the tailor has their paper pattern), the penetration into the soul of things are all steps that allow to symbolically construct the overall unit that holds everything together and to give an aura of exclusiveness to it.

The made-to-measure object is part of a symbolic world that integrates person, society and object. A world that is far away from the Baudrillardian sign world of consumerism, from which the object has disappeared, or from the Baumanian individualized world from which the individual has disappeared or from that of the solitary crowd.

3 Conclusions

From what said so far, a model that is based on holistic unity cannot be 'copied' in terms of modular fragmentation. Neither is it possible for a purely technical-functional relationship to assume a social and/or ritual function. The attempt of mass customization can be compared to the attempt once made by Taylor to analyze (in order to reproduce) the managerial work by breaking it down into parts. After thirty years of experiments and investigations he only managed to analyze the work of the most unskilled laborers.

It would be far more interesting to propose a new pattern of assignment, which could stimulate and enable clients to exercise their own creativity; however, this is hardly done in the ideal type explained above. Naturally, those businesses that wish to pursue the competitive advantage of assignment are required to: 1) undertake to train the client to be active; 2) give more 'weight' to the client by considering him/her not only a (possibly helped) selector, but a proposer, a source of possible innovations; 3) organize their structure as a service to the client; 4) allocate a commercial function also to the client-enterprise interface. In other words, the pathway to transform the nature of the business should continue until demand and supply are unified.

Bibliography

J. Baudrillard, *Per una critica dell'economia politica del segno*, Mazzotta, Milano 1974.

Z. Bauman, *Lavoro, consumismo e nuove povertà*, Città Aperta, Troina 2004.

R. Camagni, 'Cambiamento tecnologico, "Milieu" locale e reti di imprese: verso una teoria dinamica dello spazio economico', in *Economia e politica industriale*, 64, 1989.

R. H. Coase, 'The nature of the Firm', in *Economica*, 4, 1937.

D. Nelson, *Taylor e la rivoluzione manageriale. La nascita dello 'scientific management'*, Einaudi, Torino 1988.

R. Paltrinieri, *Il consumo come linguaggio*, Angeli, Milano 1998.

B. J. Pine, *Mass Customization: dal prodotto di massa all'industriale su misura. Il nuovo paradigma manageriale*, Angeli, Milano 1997.

B. J. Pine, B. Victor, A. C. Boynton, 'Porte aperte alla mass customization', in *Harvard Business Review*, (It. ed.), 4, 1994, pp. 64-77.

R. Rochefort, *Le consommateur entrepreneur*, O. Jacob, Paris 1997.

G. Simmel, *La moda*, SE, Milano, 1996.

S. Vaccà, *Scienza e tecnologia nell'economia delle imprese*, Angeli, Milano 1989.

M. R. Zorino, *Economia activa. La comunicazione, motore segreto di innovazione e mercato*, LINT, Trieste 1997.

M. R. Zorino, 'Dallo scambio tra estranei al dialogo tra produttore e consumatore', in *Metis*, 1, 2003, pp. 85-108.

M. R. Zorino, 'L'utente da consumatore di servizi a cliente-committente', in A. Porrello, *Modelli dell'utente nelle società avanzate*, IUAV, Università di Venezia 2003, pp. 77-184.

Maria Romana Zorino, Dipartimento di Psicologia Generale, Università di Padova

Yearbook of the Artificial

Published volumes

Vol. 1 Yearbook of the Artificial: Methodological Aspects and Cultural Issues, 272 pages, 2002.
Vol. 2 Yearbook of the Artificial: Models in Contemporary Sciences, 379 pages, 2004.
Vol. 3 Yearbook of the Artificial: Cultural Dimensions of the User, 249 pages, 2005.

HOMO UTENS

Identità tecnologia, cultura

a cura di
Massimo Negrotti

© 2004 Edizioni Angelo Guerini e Associati SpA, Milano, 312 pp.
ISBN 88-8107178-9 – € 24,00

La crescente diffusione del termine «utente» sembra proporre, rispetto a concetti solo apparentemente simili, quali cittadino, individuo, persona, soggetto, cliente, attore, una nuova concezione dell'uomo in seno alla società.
L'adozione di questo termine può porsi in stretta relazione con alcuni elementi culturali di fondo: la percezione diffusa di una «nuova cittadinanza», il graduale prevalere del «mercato» come riferimento ideale, l'interiorizzazione degli stili operativi delle «nuove tecnologie». Più in particolare, è plausibile ritenere che l'utente si identifichi con un soggetto intermedio tra il cittadino e il cliente, ossia tra un'entità giuridica e nominale (quella del cittadino) e una posizione meno asettica, oggetto, almeno nelle intenzioni, di una maggiore considerazione (quella del cliente).
La ricognizione, sia teorica sia empirica, sulla diffusione e i riferimenti differenziali del termine, si offre quindi come lo strumento naturale per portare alla luce tendenze di riassestamento di temi centrali nell'evoluzione culturale contemporanea.

Massimo Negrotti ha insegnato Sociologia presso l'Università di Parma e, dal 1980, in qualità di professore ordinario, Sociologia della conoscenza all'Università di Genova. Dal 1988 ricopre la cattedra di Metodologia delle scienze umane presso l'Università degli Studi di Urbino «Carlo Bo». I suoi interessi vertono sui modelli di analisi delle nuove tecnologie e, in particolare, delle tecnologie dell'artificiale. Su questi temi ha pubblicato numerosi saggi e articoli sia in Italia che all'estero tra i quali ricordiamo *Artificiale. La riproduzione della natura e le sue leggi* (Roma 2000) e *In difesa della scienza* (con G. Lanzavecchia, Milano 2002).

Kolibri 360 GmbH
Ferdinand-Jühlke-Straße 7
99095 Erfurt, Deutschland
produktsicherheit@kolibri360.de

Druck:
CPI Druckdienstleistungen GmbH
im Auftrag der
Kolibri 360 GmbH
Ferdinand-Jühlke-Str. 7
99095 Erfurt